Ecological Effects of Roads

Land Reconstruction and Management
Vol 2, 2002

The Land Reconstruction and Management Series
Series Editor: Martin J. Haigh

Vol. 1: Reclaimed Land, Erosion Control, Soils and Ecology
Vol. 2: Ecological Effects of Roads

Ecological Effects of Roads

Ian F. Spellerberg
Professor of Nature Conservation
Director
Isaac Centre for Nature Conservation
Lincoln University
Aotearoa, New Zealand

Science Publishers, Inc.
Enfield (NH), USA   Plymouth, UK

TD195.R63 S66 2002
Spellerberg, Ian F.
Ecological effects of roads

CIP data will be provided on request.

Cover Photograph Courtesy of the U.S. National Park Service.

SCIENCE PUBLISHERS, INC.
Post Office Box 699
Enfield, New Hampshire 03748
United States of America

Internet site: *http://www.scipub.net*

sales@scipub.net (marketing department)
editor@scipub.net (editorial department)
info@scipub.net (for all other enquiries)

ISBN 1-57808-198-X

© 2002, Copyright reserved

All rights reserved. No part of this publication may be reproduced, stored in a retrieval system, or transmitted in any form or by any means, electronic, mechanical, photocopying or otherwise, without the prior permission from the publisher. The request to reproduce certain material should include a statement of the purpose and extent to the reproduction.

Published by Science Publishers, Inc., Enfield, NH, USA
Printed in India.

'Land Reconstruction and Management' Series Preface

This volume examines the special features of roads, as an environment in their own right, and also their effects on the landscapes where they are created. *Ecological Effects of Roads* by Ian F. Spellerberg is the second volume in an international series of monographs and thematic collections of review and research papers based around the broad theme of *'Land Reconstruction and Management'*. The series is devoted to works that offer insights, deeper than those of the advanced textbooks, which are of immediate value to the researcher and of lasting practical value to the innovative practitioner. However, it is hoped that the series will do more than simply publish works that provide an overview of the state of play in a key area of applicable research. It also aims to address the conceptual and contextual underpinning of the applied sciences and technologies involved in practical landscape management and reconstruction. It hopes to provide a platform for new philosophies and the ideals of particular international communities, 'Schools' of applied researchers and practitioners. These include the *Land Husbandry Movement*, which seeks to transform the profession of soil and water conservation, and which provides a volume of the near future.

A key concern of the *Land Reconstruction and Management* series is the *'greening'* of the engineering effects in the environment. This theme is amply illustrated by the present volume, which considers the ubiquitous road-scape and its effects, builds upon themes introduced in *Land Reconstruction and Management's* launch volume, which also studied an entirely artificial landscape feature – the new lands created in the wake of the surface mining of coal. *Reclaimed Land: Erosion Control, Ecology and Soils,* is available from A.A. Balkema Publishers at Swets and Zeitlinger, P O Box 1675, 3000 BR Rotterdam, Netherlands: www.szp.swets.nl).

So, the series *'Land Reconstruction and Management'* deals with constructed landscapes and the management of lands that are intensively used by human societies. Its philosophy is a delicate shade of green; its core concerns are sustainability, or better self-sustainability, in landscapes that are consequent upon human activities.

The series welcomes suggestions for new contributions. If you are the keeper of a body of knowledge that fits within the remit, if you are committed to the production of knowledge useful for both the researcher and environmental management practitioner, then this series may be the right home for your work.

April 2002

Martin Haigh
(Series Editor)

Preface

The ecological effects of roads and traffic are of the same magnitude and importance as any of the topical environmental issues, for example losses in biological diversity and the ecological damage and economic costs caused by invasive and exotic species. However, this fact is not well known among environmental managers and ecologists. It is for this reason that this book was written for the Series 'Land Reconstruction and Management'.

There are many ways to reduce the ecological effects of roads and traffic and many ways to compensate for the loss of habitats due to road construction. Ecological impact assessments play an important role when planning a new road. Planting roadside verges with indigenous species can help to compensate for losses of plant communities. Perhaps the most courageous step would be a government proposal for closure of some roads in wilderness areas. Surely, in future, such proposals will become a reality.

This book commences with an overview on roads and traffic (Chapter 1) and then outlines the ecology of roads (Chapter 2). Thus the first two chapters set the scene for those that follow.

Chapter 3 deals with the biology of roads and roadside verges, the plants and animals associated with roads and whether roadside verges contribute to the conservation of nature.

Chapters 4, 5 and 6 describe the effects of roads and traffic. Habitat fragmentation, barriers and corridors are dealt with in Chapter 4 while the physical and chemical effects of roads and traffic are described in Chapter 5. The mortality of animals on roads is discussed in Chapter 6. Manner and means of reducing adverse effects are reviewed at the end of each chapter.

Methods for avoiding, remedying and mitigating the adverse effects of roads on the environment, habitats, and plants and animals are introduced in Chapter 7.

The ecology of roads in future is envisioned in Chapter 8.

Three appendices at the end of the book provide the following information: I—definition of terms; II—the process and contents of ecological impact assessments; and III—examples of ecological assessments.

Obviously then, this book is a practical guide to the general issues and their solutions. It does not provide all the answers but presents many of the challenges surrounding the ecological effects of roads and traffic.

Concomitantly, it raises the profile of the importance of these effects. Towards this objective, the ecological effects of roads are sometimes discussed in the broader context of road culture and road users.

Over the last decade of researching the ecological effects of roads and traffic, I compiled a collection of quotes that seem to demonstrate an evolution of thought and a changing culture towards roads and their benefits and costs.

> Planning for Man and Motor (1964)
> There is a tendency for manufactured features…to be given a higher priority than areas of nature conservation value (1990)
> The road to recovery (1992)
> A road sold as a nature reserve (1993)
> Birds object to road roar (1995)
> Office bypasses new road rules (1995)
> One more for the road (1995)
> Smooth road skirts tender-toed Kauris thanks to green 'boffin' (1996)
> The road transport system is one of the largest areas of environmental impact (1997)
> The AA and RAC are not just breakdown companies (1997)
> The killer gulleypots (1997)
> Road toad. It needs a ladder (1997)
> Traffic reduction gets green light (1997)
> Road reform and the environment (1998)
> A river of traffic and its tributaries (1998)
> Business groups are campaigning for more motorways (1999)
> Transport snarlup (1999)
> Road rage (2000)
> The lawless roads (2000)
> Road enmeshed in politics (2001)
> I have given three commitments: Roads, roads and roods. (2002)

Many colleagues, friends and family members assisted in assemblying the material for this book, as well as in preparation of the manuscript. I gratefully thank the following who made a significant contribution: Alan Ambury, Helena Bender, Cas Besselink, Helen Byron, Marnie Criley, Barry Fox, Martin Haigh, Nicholas Martin, John Sawyer, A. P. Singh, Karen Spellerberg, Kathryn Spellerberg, Jane Swift, and Fritz Voelk. I would like to acknowledge the contribution made to this area of research by the Late Clare Washington. Clare, like me, had a passion for research on the ecology of roads. I also thank those publishers and authors who kindly allowed me to include material from their work. Insofar as possible, copyright permission was sought in all cases.

Ian F. Spellerberg

Contents

Series Preface	*v*
Preface	*vii*
1. Roads and Traffic	1
2. Scoping the Ecology of Roads and Traffic	19
3. Biology of Roads and Roadside Verges	45
4. Habitat Fragmentation, Barriers and Corridors	67
5. Physical and Chemical Effects of Roads and Traffic	98
6. Road Kills: Animal Mortality on Roads	118
7. Reducing Adverse Effects	135
8. The Ecology of Roads in the Future	164
Appendix I: Definitions	168
Appendix II: Environmental and Ecological Impact Assessments	174
Appendix III: Examples of Environmental Impact Assessments	192
References	225
Index	249

1

Roads and Traffic

INTRODUCTION TO ENVIRONMENTAL EFFECTS OF ROADS

Visitors from outer space would on approaching the planet Earth, make out the continents, then large islands and natural features such as the Great Barrier Reef off the north-east coast of Australia. On circling the Earth away from the sun, they would be drawn to the sparkling lights of cities and busy motorways. With the sun behind them, they might detect the Great Wall of China and almost certainly canals, railway lines and road networks. The extent of forest clearings due to the spread of new roads far from any cities (Fig. 1.1) might puzzle them.

Throughout the world, roads and traffic have become a permanent part of our physical, social and cultural environment. From Antarctica to the Arctic, from the coasts to the mountains, from tunnels under the sea to tunnels in the bowels of mountains, roads and vehicles are everywhere. Most roads are on land and have a gravelled or sealed surface in polar regions, there are roads on sea ice. In Canada, temporary logging roads are constructed out of the ice of frozen rivers (Fig. 1.2). Few are the places in the world where one can escape the sight of roads and the sound of vehicular traffic. News about roads and vehicles appears in the media every day.

Roads have long been important for development and prosperity. Many organisations support this view today and argue strongly in favour of more roads, based on the needs of road users. Indeed, in Britain, the long-established Automobile Association (AA) and the Royal Automobile Club (RAC) advertise that they are not just vehicular breakdown services, but powerful lobbyists for the road-building industry. Interestingly, another motoring organisation, Environmental Transport Association (ETA) (www.eta.co.uk), launched only in 1990, has grown rapidly. It claims to be an ethical alternative to other motoring organisations and actively campaigns for a sustainable transport system.

Sustainable transportation seems to be on the environmental agenda for many countires. For example, in Canada, the National Round Table on the Environment and the Economy established a programme on sustainable

Fig. 1.1. Satellite photograph of deforestation in Rodonia, Brazil. The activities in this area are described by Nelson and Holben (1986). The white lines (4 km apart) are colonisation clearings that extend from new roads. In Central America, Kaimowitz (1996) identified road construction (along with favourable markets for livestock products, subsidised credit, land tenure policies, policies that reduced timber values and levels of political violence while improving cattle husbandry) as one of the processes that resulted in replacement of forest by pasture. The same processes may have been applied in Brazil (photo, Brent Holbern).

Fig. 1.2. Examples showing the diversity of roads.

Top left: A Roman road (photo, Myfanwy Spellerberg). Top right: end of the road! in Brazil (photo, WWF photolibrary). Middle left: An ice bridge or ice road built on Williston Lake, BC, Canada (photo, Ern Hegglun). Middle right: Winchester bypass in southern England (photo, Ian Spellerberg). Bottom left: Temporary forest road (photo, Ian Spellerberg).

transportation. A report was released in 1997 and recommendations agreed in the areas of education and awareness and reduction in environmental impacts (especially vehicular emissions).

Roads have impacted on the past and will impact on the future. Their contribution in the past has been so great, in fact, that some would say roads helped to shape our history while modifying our environment. The long straight Roman roads in England remain a landscape feature, albeit in the form of modern roads. Many roads were built to expand empires, establish trade routes and provide access to new natural resources. Whether for reasons of war, politics, or trade, roads have and will shape our future.

Roads of the future will differ in design, however, being built and managed with nature in mind. Road routes will be laid out to cause the least impact on nature, provide more wildlife crossings and encompass better landscaping of roadsides using plants native to the area.

GROWTH OF ROADS

Some road statistics are given in Table 1.1. The extent of road networks and the rate of growth in roads can be expressed in various ways, such as total length, per cent increase, policies advocating road building, and extent of habitat lost to roads.

In many countries there has been, and continues to be, tremendous development in road construction. China has one of the largest road expansion projects in the world. Mexico has been attempting to double its length of toll roads. In Asia, the Asean Highway Project is planning development of the Asean highway network. The objective, as stated in 1998, is to connect transportation of high potential areas of all member countries. The network comprises 23 routes. The Asean highway development programme is just one of several UNESCAP projects (United Nations Economic and Social Commission for Asia and the Pacific).

In 2000, the World Bank announced that it would assist Kerala, India in highway development. Some 1800 km of highways and major roads are to be constructed in Kerala (the government is aiming for 2800 km). Part of the rationale for this development has been complaints from tourists and local people that travel between the airport and the city took too long.

The World Bank (together with the Asian Development Bank, the Swedish National Road Administration and the Overseas Development Administration in the UK) is funding modern tools for highway development and management. The International Study of Highway Development and Management was initiated in 1993 to produce improved tools for the development of economic and technical strategies in the road sector.

In Britain, the overall road stock increased by 5.4% from 1978 to 1988. In 1998, the White Paper 'Roads for Prosperity' (UK: Department of Transport, 1989) outlined proposals for a large increase in motorway and trunk road

Table 1.1: Road statistics: lengths and area of roads in selected countries

Australia
Total length of roads approximately 870,000 km. Total length of rural roads about 780,000 km. Area of rural roads about 31,200 km^2 (assuming an average width of 40 m) (from Straker, 1998).

China
China is one of the countries with the highest growth rate of road development in the world. Road building is already extensive and more is planned to cater for the 15.8 million motor vehicles. From 1996-2000, a total of 13,000 km of motorways was constructed and an additional 9000 km is to be built by 2005, bringing the total to 22,000 km. In the same period, 240,000 km of highways were built, bringing the grand total of expressway length to about 1.4 million km.

Great Britain
Total length of roads approximately 371,914 km. Between 1989 and 1999, there was a 4.1% increase. The total area of land taken up by roads in 1999 was about 3200 km^2 (constituting 1.4% of the total land area).

India
Total length of roads about 2.03 million km, of which about half are sealed. Many plans are afoot for road improvements but all are hampered by limited resources. According to Living Media India Ltd., India is approaching a road famine.

New Zealand
Total length of maintained roads 92,000 km (used by 2.3 million vehicles). Total area taken up by roads about 887 km^2 (about 0.33% of the total land area). Estimated area of rural road verges 66 km^2.

The Netherlands
Road infrastructure takes up approximately 1.5% of the land area (Bohemen, pers. comm.). The area of roadside habitats is considerable and estimated to be about 60,000 ha, or about 1.5-2.1% of the total land area of the country (Schaffers, 2000).

USA
Over 6.2 million km of public roads are used by 200 million vehicles (Hamilton and Harrison, 1991). These roads with roadsides cover about 1.0% of the land area. More than 20 million acres constitute highway right-of-ways (Leedy, 1975). Based on Forman's calculations (Forman and Deblinger, 2000), 19% of the land area is directly affected ecologically by this system of public roads. Roadside vegetation is said to comprise over 6.2 million ha (Mowbray, 1968).

construction in order to reduce road congestion and to cater for a 100% increment in road traffic by 2025.

One example of extent and rate of habitat loss caused by road construction is found in The Netherlands. The loss of habitat during 1980-1993 resulting from construction of rural sealed roads was 180 km^2, constituting an annual loss of 0.04% of rural land (Cuperus et al., 1999).

STRUCTURES ASSOCIATED WITH ROADS

Besides implications of the geographical location of roads, there are also implications in terms of structures associated with roads, such as rest areas, car parks telegraph poles and wires, fences, bridges and tunnels. Some wildlife use tunnels as habitats and birds may use overhead wires and electricity poles and transformers for nesting or roosting (Chap. 3, Fig. 3.2b).

Fences are common structures alongside roads, often used to prevent access of domestic animals to roads but precluding wildlife as well. Their design and location have been the focus of many studies. For example, Feldhamer et al. (1986) reported the effects of interstate highway fencing on white-tailed deer activity in Pennsylvania, USA (also see Chap. 4). They found that a 2.7-m fence reduced the number of deer on the right-of-way compared to a 2.2-m fence, but was not effective in reducing the number of road kills. Habitat, topography, width and kind of roads, and traffic density are some factors that must be considered while assessing the effectiveness of roadside fences for wildlife (see Chap. 4).

ENGINEERING ACHIEVEMENTS

Roads have been a major focus of attention for engineers and for transport and road research laboratories. The extent to which engineers have been involved with roads is perhaps matched only by the rich variety of literature on road construction. Some of this literature is notable for absence of any reference to ecology. Even in more recent publications where ecology is mentioned, the reference is simple and naïve, often referring merely to the noise effects on birds.

In *The Construction of Roads and Pavements* published in 1916, Thomas Radford, not surprisingly, makes no mention of ecology or ecological effects, because ecology had only just begun to emerge as a distinct discipline. (The British Ecological Society was established in 1913 and the Ecological Society of America followed in 1916.) More recently, in 1981, the seminal book *Environmental Impact of Roads and Traffic* by L. H. Watkins makes no reference to ecology in the contents but does refer to roadside plantings—with emphasis given to landscape rather than to ecology and ecological effects.

Roads have been constructed in some of the most montane regions of the world. In the Himalayas, for example, roads seem to cling to steep mountain sides. Haigh (1983) presents a vivid picture:

> The Karakoram Highway is a strategic, all-weather, blacktop motor road, and follows the contours as it clings to the edge of the Indus Gorge to Gilgit, then up to the Khunjerab Pass (4935 m) and China. The Highway, whatever its military and political significance, is a monumental engineering achievement. The road crawls across cliff faces, traverses slopes steeper than 80° at one point for each kilometre block of the road,

edges across unstable landslip zones, past the foot of a dangerously active glacier, and along the edge of screes of loose debris standing at 37° and reaching 300 m above the road. A survey of the hazards along the road included steep rock slopes with incipient slab failures, overhanging slopes with rocks loosened by dilation joints and blast fractures, missing culverts over ephemeral gullies.

Hazard assessment in other montane regions of the world has helped to establish a spectacular way of overcoming the frequent damage to roads caused by landslides and washouts. Part of Highway 73, west of Arthur's Pass (924 m) in the Southern Alps of New Zealand, descends abruptly into the Otira Gorge. The cost of frequent repairs to this section of the road and the implications of diverting traffic to alternative routes contributed to the decision to build a spectacular viaduct extending down the Gorge.

A WORLD CAR CRISIS?

When first introduced to roads, vehicles such as cars were not well received. In the UK, the Locomotive Act of 1865 required any car in motion on a road to be attended by at least three persons, one of whom had to walk some metres ahead carrying a red flag by day or a red lamp at night. One hundred years of motoring saw rapid changes, exponential growth in number of cars and an infrastructure to support road transport. Now there is a road transport crisis.

The exponential growth in number of cars on roads is a worldwide phenomenon. The number of motor vehicles is huge. In China, for example, there are 15.8 million motor vehicles and the increase in numbers is outpacing the capacity of roads to support the traffic. Once a rare sight, a luxury, a romantic experience, cars and traffic density have now become a major environmental issue. The increase in problems brought about by the age of the car and by transportation in general has been explored in many reports and texts (e.g. Fletcher and McMichael, 1997; Gordon, 1991; Greene et al., 1997; Zuckermann, 1991). The World Car Crisis is well summarised by Zuckermann (1991):

— 500 million vehicles on the road,
— mounting traffic congestion,
— pollution with implications for health and climate change,
— dependence on fossil fuels,
— 250,000 traffic deaths per year, and
— 50 million new vehicles each year.

A ROAD CULTURE

Roads, streets and vehicles, especially cars, form an integral part of the lives of millions of people in most parts of the world. Not surprisingly, a language and a culture have been built on the common associations with and uses of roads, streets and cars (Table 1.2).

Table 1.2: Road and street expressions coined in the English language

Roads
    One for the road
    Road hog
    Road rage
    Road show
    Road test
    Roadie
    Roadster

Streets
    On the streets
    Streets ahead
    Street-walker
    Up one's street
    Street-car

Associated with the high dependence on road vehicles and the continual growth in motorised traffic are some very powerful organisations and lobby groups. The vehicular industry is a very powerful lobbyist as are truck driver unions and other major road users. Attempts to reduce vehicular road use have often been defeated by such powerful forces. Truck driver unions have defied governments. In Japan, truck drivers are known as 'Tarmac Warriors', having their own social hierarchy and structure. In northern India, truck drivers adorn their cabs with protective deities, among which the most popular are Durga (Goddess of Battles) and Lord Hanuman, who carried a mountain from Sri Lanka for his master Rama.

Cultural links and myths about roads appeared when roads were first built (e.g. 'The Road to Damascus'). Today, some highways in the USA have a near-mythical status relating to the odyssey and sea voyages of ancient times. Roads and vehicles and travel by car have all contributed to book titles (e.g. Greene's *The Lawless Roads*), films (Table 1.3), poems and songs. For example, Highway 61 in the USA is referred to as 'the Blues Highway' because it leads directly to the place of origin of the blues. Walt Whitman wrote the 'Song of the Open Road' ('Afoot and light-hearted, I take to the open road, healthy, free, the world after me').

Roads have become a symbol of a way of life. In their book *Landscape Narratives*, Potteiger and Purinton (1998) demonstrate very clearly the extent to which roads and traffic have contributed to a way of life. They epitomise it in this brief excerpt referring to Highway 61: 'With its themes of mobility, escape and cultural crossover into rock-and-roll, the story of the blues merges with the story of the road.'

Table 1.3: Films with reference to roads (based on Potteiger and Purinton, 1998)

The Grapes of Wrath (1940)	Something Wild (1986)
Detour (1946)	Candy Mountain (1987)
Gun Crazy (1949)	Mystery Train (1989)
La Strada (1954)	Wild at Heart (1990)
The Wild One (1954)	Highway 61 (1991)
Pierre le Fou (1965)	My Own Private Idaho (1991)
Bonnie and Clyde (1967)	Roadside Prophets (1992)
Weekend (1967)	Leaving Normal (1995)
Easy Rider (1969)	Mad Max (1995)
Badlands (1973)	To Wong Foo, Thanks for Everything, Julie Newman (1995)
Wrong Moves (1978)	
Roadie (1980)	Feeling Minnesota (1996)
The Road Warrior (1982)	Crash (1997)
Stranger than Paradise (1985)	Endless Highway (1997)
Down by Law (1986)	

There is a humorous side to most things and that includes the extent and ramification of roads (Box 1.1). The range of cartoons prompted by roads and traffic could probably fill a book.

ROAD POLICIES AND LEGISLATION

For decades throughout the world there has been a repeated cycle of road building, growth in traffic, traffic congestion and more road building. The solution to growth in traffic has typically been to build more roads. Strong underlying political pressures have shaped road policy. In Europe, road networks have been expanded, particularly in the 1960s and 1970s, to cater to increased road traffic. In the UK, between 1979 and 1997, the 18 years of Conservative Governments led to private ownership of transport services. Subsequently, the Labour Government has adopted an integrated transport policy with greater regulation.

By 1980, however, several countries and organisations in various parts of the world had come to recognise that a transportation crisis existed. Thus growth of road transport and environmental impacts became major issues for consideration.

In 1989, the European Conference of Ministers of Transport (ECMT) and the Organisation for Economic Co-operation and Development (OECD) met to discuss the mounting issues of transport and the environment. While recognising the major and positive role in economic life played by transport, such negative effects as accidents, congestion, pollution, noise, energy consumption, land consumption, and destruction of neighbourhoods, farms and wildlife habitats had to be faced squarely. Ironically, although 'wildlife habitats' appear in the introductory pages of the subsequent report, there is no mention of them in the recommendations!

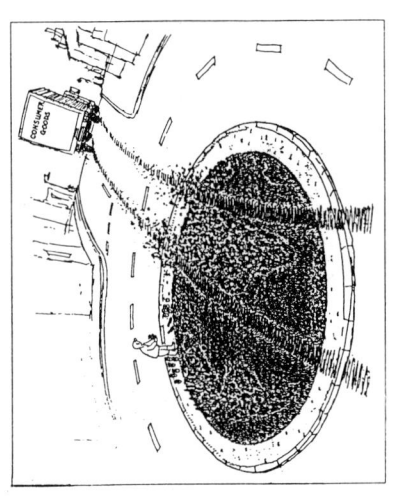

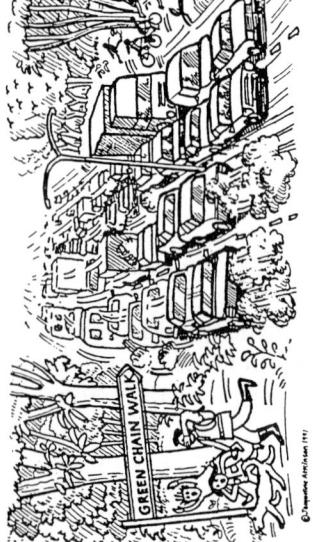

Box 1.1: Some road and traffic cartoons. The 'Green chain walk' is by Jacqueline Atkinson and the other cartoons by Chris Madden. The 'Lorry and the roundabout', published in BBC Wildlife 1994, was the winner of the Guardian Prize.

Meanwhile, the OECD had established a road research programme in 1967, which had now become road transport research with 40 members participating. The 30th Anniversary Report of the OECD published in 1997 recognised environment-transport interactions as one of eight themes. However, the approach to impacts on the environment was very human orientated (OECD, 1997).

Sustainable transportation had now become a paramount issue. A framework for a sustainable transport policy was included in the 1994 Sustainable Development Strategy for the UK. It included the following:

- Strike the right balance between the ability of transport to serve economic development and concomitantly protect the environment and sustain the future quality of life.
- Effect measures that reduce the environmental impact of transport and influence the rate of traffic growth.
- Ensure that users pay the full social and environmental costs of their transport decisions, thereby improving the overall efficiency of those decisions for the economy as a whole and effecting environmental benefits.

Within this framework, the UK Government acknowledged that it would have to address several matters, including the following:

- Influence the rate of traffic growth and provide a framework for individual choice, thereby enabling environmental objectives to be realised.
- Extend public awareness/understanding of environmental impacts caused by polluting emissions from vehicular traffic.

There are currently pressures in other countries also to restrain the growth of road transport. For example, in the Netherlands, the Government has desisted from encouraging growth in traffic and further development of the road network. In the UK too, after many years of pressure to review policies on road development, the Government is beginning to shun the 'more roads' solution.

BENEFITS AND COSTS OF TRANSPORTATION

Texts on costs and benefits of transportation and human health as aspects of transportation include O'Flaherty (1997), Greene et al. (1997), Fletcher and McMichael (1997) and Glaister et al. (1998). Maddison et al. (1996) have considered and debated the true costs of road transport.

The methodology for identifying costs and benefits of a road project was first proposed back in the mid-nineteenth century. A cost-benefit analysis (CBA) or COBA (a benefit-cost analysis in the USA) is a structured method or procedure for comparing alternatives. In short, a CBA is used in decision making to improve social outcomes. It provides a basis for acceptance or rejection of a project by identifying the 'best' alternative.

The benefits of a new road could include the following:
— contribution to economic growth,
— reduced journey time,
— reduced transportation cost, and/or
— new employment opportunities.

Thus the economic and social benefits of roads are clear. They support supply chain networks and means of transportation communication between centres of human populations. By improving social and economic welfare, they have provided and continue to provide huge benefits to many generations of people throughout the world. Strong, statistically positive correlations have been shown between road density and per capita gross national product in several countries (Wilkie et al., 2000). More roads mean more wealth. The benefits can be quantified in monetary terms (see Box 1.2).

Box 1.2: Benefits and Costs of Transportation in the USA in 1994 (from Greene et al., 1997)

Benefits include (in addition to individual services)
  4.4 trillion passenger miles of travel
  More than 3.5 trillion ton miles of freight movement
  Total expenditure on transportation services $1 trillion

Costs include
  43,881 lives lost
  Major cause of air pollution
  Vehicles generated 0.75 of CO emissions
  Consumption of 65% of petroleum products
  In urban environment, noise and social issues are created

The costs of building a road and its subsequent maintenance generally include the following:

— capital expenditure (construction costs);
— costs of raising the capital;
— maintenance costs;
— labour;
— professional fees for consultants;
— materials;
— opportunal costs (with limited financial resources, road costs may preclude construction of a local amenity)
— insurance;
— costs of accidents;
— methods for overcoming social impacts (such as a major road dividing the community of a town);
— methods to minimise pollution, noise etc.;
— crop losses attributable to vehicular pollution;

— costs to mitigate greenhouse gas emissions;
— oil spill clean-up costs;
— compulsory purchase of properties;
— landscaping;
— land swaps for conservation; and
— avoidance, mitigation or compensation of ecological effects.

This list is far more extensive than that given by O'Flaherty (1997), in which only construction costs, maintenance costs and costs of delays are mentioned.

Many of the costs of a new road are readily quantified in monetary terms. As shown above (Box 1.2), in the USA alone in 1994, the benefits of transportation were vast but so too were the costs.

But what about social costs? Heavy road traffic may disrupt communities and have impacts on mental health. Some studies have shown that as traffic volume increases, social intercourse declines. In general, heavy traffic changes human domestic behaviour: front gardens and front parts of houses may be used less often. The effects of traffic noise and other forms of pollution have been widely recorded. Even the detrimental effects of roads and traffic, such as human mortality on roads, are calculated in monetary terms. The cost of a human life (road mortality) has been valued in various countries from hundreds of thousands to millions of dollars.

The cost of building and maintaining a road can well extend beyond the physical boundaries of the road. Very often, a huge infrastructure is required for roads and road transport. Industries and governmental agencies are devoted to road planning, policy, construction and maintenance.

Environmental costs can be huge and long lasting (Table 1.4). For example, the environmental costs (pollution, climate change, noise and vibration) of the transport system in the UK for 1994/1995 were 4.6-12.9 billion pounds (UK: Royal Commission on Environmental Pollution, 1994).

There are methods used to quantify environmental costs. These methods help to identify different kinds of benefits and values (including the value of commodities and services; bequest values of habitats, which may benefit future generations and existence values—the benefits people receive by simply knowing that the natural habitat exists). One example of economic values associated with recreational areas in which there are roads and those in which there are no roads has been well described by Swanson and Loomis (1998). They identify the economic benefits of federal land management in the USA and the role played by cost-benefit analysis.

Despite these analytical techniques, the full environmental costs of roads are all too often ignored or avoided, partly because economic values are perceived to be more important than environmental values and partly because of the difficulty of assigning monetary values to the adverse ecological effects. Habitat fragmentation, loss of plant populations and habitat degradation can be quantified in monetary terms, but it is difficult to put a price on the loss of nature.

Table 1.4: Estimates of costs (in billions of pounds per year) of the environmental transport system in the UK[a, b]

	Range[c]	
	Lower end	Upper end
Air pollution	2.4	6.0
Climate change	1.8 (4.6)	3.6 (12.9)
Noise and vibration	1.2	5.4
Accidents	5.5 (5.4)	5.5 (5.4)
Total quantified environmental costs	10.9 (10.0)	20.5 (18.3)

[a]From UK: Royal Commission on Environmental Pollution, 1994.
[b]Costs attributable to road transport are shown in parentheses.
[c]Among environmental costs for which a monetary value could not be estimated and which are therefore not shown in the table are:

— land losses
— loss of access to land
— visual intrusion
— severance of communities
— loss or disruption of habitats.

The scale of costs and benefits of transportation issues is now such that mounting costs are set against the benefits. In southern Belize for example, road building in areas of poor agricultural land and low population density seems to have resulted in a 'lose-lose' situation because of low economic returns and habitat fragmentation (Chomitz and Gray, 1996).

Few motorists know that they barely contribute in monetary terms to the true costs of motoring. There are many unintended consequences and undesirable impacts of transportation on the environment.

The cost of transportation is one matter, but if all environmental costs were to be included in road-building projects, then who would pay? This question, although very topical as the costs of transportation soar, is outside the scope of this book.

CONFLICTS ABOUT ROADS AND ROAD PROPOSALS

Most people take roads and traffic for granted. They are part of our culture and our way of life. Roads are considered beneficial despite the many problems—traffic congestion, air pollution and damage to buildings where traffic exceeds certain densities. Traffic accidents account for many deaths and injuries but some would say that overall the benefits outweight the costs (both in monetary and social, economic and even environmental terms). However, the number of people drawing attention to problems of traffic congestion and environmental pollution from traffic is rising. 'More roads to ease increasing traffic density' is a cry challenged more often these days.

When it is said that roads have no negative effects, perhaps what is meant is that the positive effects outweigh the negative. The view that roads are entirely beneficial, with nothing but positive effects, is rather narrow, misinformed and short-sighted.

Roads and traffic produce many effects on the environment, the scale, magnitude and implications of which are rarely acknowledged. The area of land taken up by roads and extent of roads (and associated impervious surfaces such as car parks) partially accounts for about 25% of the world's land surface presently under sealed surfaces or concrete or under the plough. The ecological effects of this fact have only recently been recognised and measures to reduce the negative effects are fortunately being introduced.

Roads have been the cause of much public concern for many decades now. Road transport projects can generate strong passions—whether for or against them. Concern has been expressed over traffic congestion. Public demonstrations have supported construction of new roads and improved transport systems. Powerful organisations insist on further road development and road lobby groups are fearsome advocates of the same (Hamer, 1987).

Campaigns against roads have become more common. In the UK for example, strong opposition to new road schemes was particularly evidenced in the 1980s and 1990s. One such proposed road scheme was the infamous Winchester bypass in southern England. Opponents proposed that a tunnel be built but a decision was eventually taken to cut through the chalk hill. In doing so, conservation and archaeological sites were destroyed. The scale of swaths cut through the hill has been shown in Fig. 1.2.

Judging by the content of news article headlines (Box 1.3), the planning of roads raises some questions about the possible implications for the natural environment and the plants and animals that constitute it. Questions of priorities come into play as do the value of nature and ethics. It is difficult, for example, to decide fairly and rationally between damaging a wetland versus damaging an ancient woodland.

There are many studies showing that more roads create more traffic. Not surprisingly therefore, some people believe that new roads must not be constructed. Campaigns to save wildlife areas are increasingly more common. One notable example was the campaign to save the Oxleas wood near London. The history and rationale of the battle against the east London River crossing is well documented in David Black's report (1993). The illustration by Jacqueline Atkinson (Fig. 1.3) underscores the feelings of those who campaigned to save the Oxleas wood.

Some people are more and more often going to extraordinary lengths to oppose new road schemes or proposals to modify existing roads. The issue of road construction can sometimes be very emotive and groups have been established to contravene proposed road programmes. Perhaps some of these objections could be termed the NIMBY attitude (Not In My Backyard!) or BANANA behaviour (Build Absolutely Nothing Anywhere Near Anyone). However, without doubt, the objections to roads in the 1980s and 1990s were the result of concern about the loss of and damage to nature.

Box 1.3:

Examples of road concerns as expressed in some newspapers. 'Anti-road army' (*The Daily Telegraph*, UK, 1995); 'Paying the price for roads' (*The Guardian*, UK, 1992); 'Wildlife Dept. objects' (*The Star*, Sarawak, 1997); 'Protesters dig in' (*The Times*, UK, 1995); 'Majority would prevent road' (*New Zealand Herald*, 1995); 'Worst route' (*New Zealand Herald*, 1994).

But there is more to roads than the loss of and damage to nature. The greatest environmental impact of roads is their slow but steady incremental growth and expansion (Box. 1.4).

WHAT THIS BOOK IS ABOUT (AND NOT ABOUT)

Roads are the main conduits for traffic although other conduits for different kinds of traffic do exist, namely canals and railways. The ecological effects of

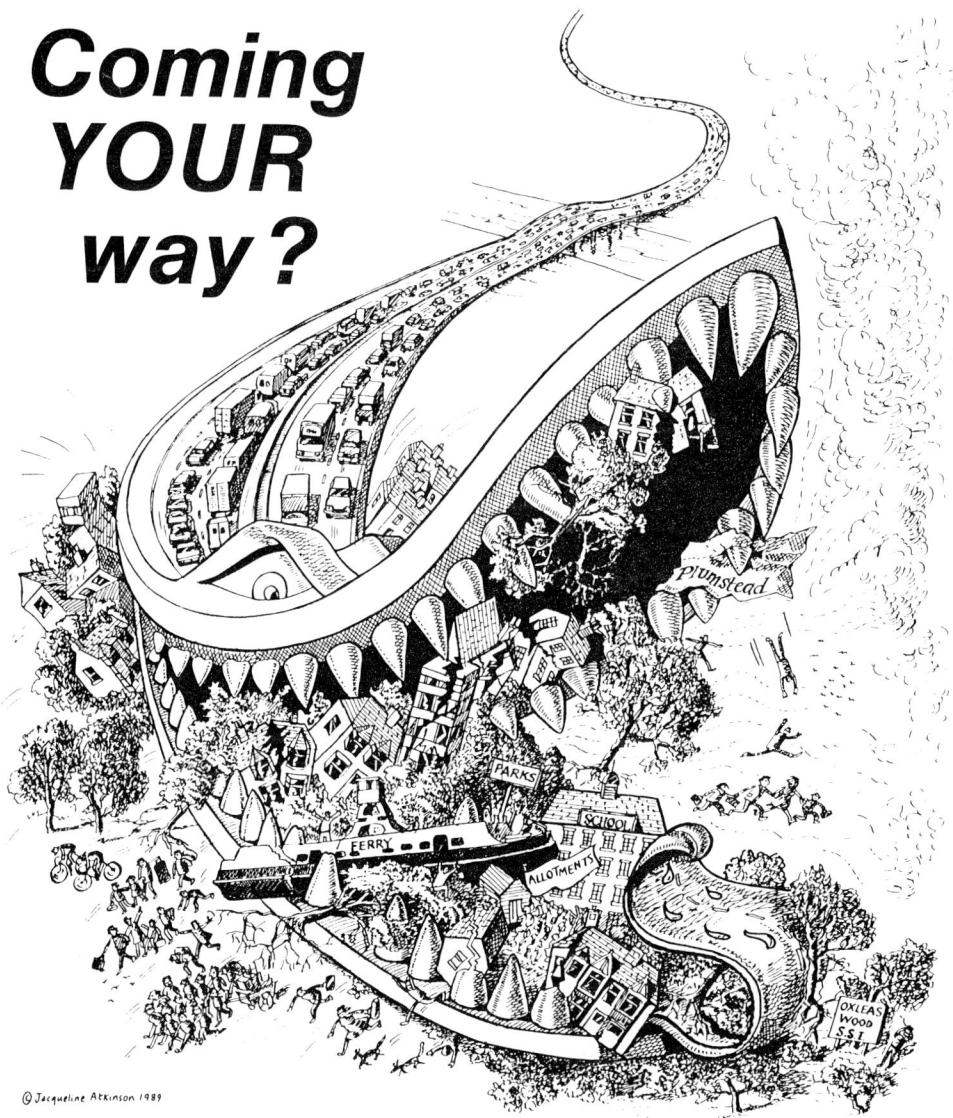

EAST LONDON RIVER CROSSING

Fig. 1.3. Cartoon by Jacqueline Atkinson.

canals and railways are not considered here but nevertheless some aspects of roadside ecological restoration and mitigation of effects would apply to linear features in the landscape other than roads.

This book deals with the ecology of roads and vehicles and the ways whereby adverse effects on the environment and plants and animals can be

> **Box 1.4: Roads Beget More Roads**
>
> It is as though a road were some kind of living creature. A living creature which begins life as a long thin organism that then sprouts tentacles. These tentacles slowly extend outwards across the landscape, their sinuous movements changing habitats, destroying species and infecting some areas with invasive species. What starts as a single, narrow swathe cutting across the countryside ultimately and inevitably will branch and become larger, destroying larger and larger areas around and beyond the road. The greatest ecological impact of a road is its cumulative effect over time. The slow pace of accumulated effects is barely noticed in the short memories of human beings because they all too readily accept slow changes in their surroundings.

avoided, remedied or mitigated. Hence the purpose of this book is to review the effects of roads and vehicles on nature and the environment and to assess the ways of avoiding, reducing and mitigating these unfavourable effects. Without doubt, the incremental impacts of roads and motorised traffic have contributed either directly or indirectly to the loss of plants, animals and biological communities in all parts of the world, even in the remote and isolated continent of Antarctica.

The environmental effects of roads are many (see Chap. 2) and may extend beyond roads. Off-road vehicles and how they affect the ecology of dunes and wilderness areas constitute a separate subject, mentioned only briefly in a later chapter. The infrastructure for vehicles when off the road (parking facilities) and the industry associated with vehicles or their parts no longer in use (wrecker yards, tyre dumps, oil disposal) are not discussed at all. Research on the environmental impact of car scrapping has centred on the influence of a car's lifetime on the life cycle of NO_x and VOC emissions (van Wee et al., 2000).

Tyre dumps have created huge and far-reaching environmental impacts. The used tyre trade proved a haven for mosquitoes, which bred in the containers (Reiter and Sprenger, 1987). Efforts have been on to address the issues of both tyre and oil disposal, including processes whereby scrap tyres and waste oils could be converted into a material suitable for use as a bitumen modifier (Ksaibati et al., 1995).

Space does not permit technical details of the methods used for reducing or compensating for the ecological effects of roads and traffic because these methods necessarily vary from region to region depending on local conditions and geography. Furthermore, the technical details of the methods commonly used require lengthy descriptions. Hence this book is confined to the general issues of planting roadside verges. Details can be found in the References provided at the end of the book.

2

Scoping the Ecology of Roads and Traffic

INTRODUCTION

This chapter commences with a brief account of research and other interests in the environmental effects (excluding ecological) of roads and traffic. The ecological effects are treated in a separate section subsequently. So the primary purpose of this chapter is to scope the general environmental and ecological effects of roads and traffic.

RESEARCH, DATABASES, REPORTS, ORGANISATIONS AND CONFERENCES

Interest in the ecology and natural history of roads is not new. Observations of the effects of roads on roadside ecology have been published since the first half of the 1900s (e.g. Simmons, 1938; Dickerson, 1939). Nevertheless, research on the ecological effects of roads was rather scant until the 1990s, centring only more recently on the ecology of roadsides and the effects of habitat fragmentation caused by roads. Earlier twentieth century publications were more in the form of observations than research *per se*. Attention focused on the mortality of animals on roads, which over the decades caused sufficient concern to label roads as 'long, narrow slaughterhouses' (Schullery, 1987), while other authors extolled the benefits of roadsides for some wildlife.

Literature (some lacking bibliographies) reviewing the effects of roads on nature and occasionally including information on mitigation of such effects has, as mentioned above, been published in several countries (Table 2.1). Many countries have an interest in the ecology of roads, but most of the literature published on this topic originated in the USA, followed by the UK.

Some road literature databases are also available.
Transportation Research Board Publications (USA)
An annotated bibliography on the ecological and environmental effects of highways.

Road-RIP Roads Bibliographic Database

Compiled by the Wildlands Centre for Preventing Roads (CPR), USA. Includes about 6000 citations relating to the ecological effects of roads. Topics range

Table 2.1: Reports and reviews on the effects of roads and traffic

a) Environmental effects

Scheidt, M. 1971. Environmental effects of highways, pp. 419-492. In Detwyler, T. R. (ed.). *Man's Impact on Environment.* McGraw Hill Book Co., NY.

TEST. 1991. Wrong side of the tracks. Impacts of road and rail transport on the environment: a basis for discussion. *TEST Report* 100, London.

Watkins, L. H. 1981. *Environmental Impact of Roads and Traffic.* Applied Science Publ., London.

b) Ecological effects

1. *Books and reports (with reviews) on effects of roads and traffic*

Aanen, P. et al., 1991. *Nature Engineering and Civil Engineering Works.* Pudoc, Wageningen.

A collection of papers on the relation between nature engineering and civil engineering works. This book includes many examples of mitigation methods.

Edmunds, J. 1995. *Head-on Collision 1995.* Wildlife and Roads Report. Cheshire, Cumbria and Lancashire Wildlife Trusts.

Threats to important wildlife sites from road developments are described with reference to the UK—Cumbria, Lancashire, Merseyside, Greater Manchester and Cheshire.

Environmental Resources Management. 1996. The significance of secondary effects from roads and road transport on nature conservation. *English Nature Research Reports.* 178. English Nature, Peterborough.

A comprehensive report for English Nature by Environmental Resources Management.

Kavtaradze, D. N. et al., 1999. *Automobile Roads in Ecological Systems.* Moscow State Univ., Moscow (in Russian).

Brings together research reports on the impacts of pollutants, impacts on plant, animals and human beings, and discusses road ecosystems.

Noss, R. 1995. The ecological effects of roads or the road to destruction. Unpublished.

Not very objective; roads are considered bad from the start.

Ramsay, D. (ed.). 1994. *Roads and Nature Conservation. Guidance on Impacts, Mitigation and Enhancement.* English Nature, Peterborough.

Comprehensive report by Penny Anderson Associates.

2. *Reviews in anthologies or journal articles*

Conservation Biology. February 2000.

Special section on ecological effects of roads. Includes nine papers on various aspects of ecological effects of roads in North America.

Andrews, A. 1990. Fragmentation of habitat by roads and utility corridors: a review. *Australian Zoologist,* 26: 130-141.

Very good review of the literature, especially with reference to habitat fragmentation.

Atkinson, R. B. and Cairns, J. 1992. Ecological risks of highways. Advances in *Modern Environmental Toxicology,* 20: 237-262.

Comprehensive review of the ecological effects ranging from habitat fragmentation through genetic risks to ecosystem stress.

Contd.

Table 2.1: (Contd.)

Bennett, A. F. 1991. Roads, roadsides and wildlife conservation: a review. In Saunders. D.A. and Hobbs, R. J. (eds.). *Nature Conservation*, vol. 2: The Role of Corridors. Surrey Beatty & Sons.
  Very good review of the literature.
Forman, R. T. T. 1995. Land mosaics, the ecology of landscapes and regions. In: *Corridor Attributes*, Cambridge Univ. Press, Cambridge.
  Good review of roads as linear landscape features.
Gilbert, O. L. 1989. *The Ecology of Urban Habitats*. Chapman and Hall, London. Chapter 9: Roads. Good analysis of the literature.
Leedy, D. L. 1978. Highways and wildlife: implications for management, pp. 364-383. In Classification, Inventory, and Analysis of Fish and Wildlife Habitat. *Proc. National Symp*. Biological Services Program. Fish and Wildlife Service, U.S. Dept. Interior, USA.
  An overview of the good and bad aspects of impacts of effects of roads on nature.
Schonewald-Cox, C. and Buechner, M. 1992. Park protection and public roads. In Fiedler, P. L. and Jain, S. K. (eds.). *Conservation Biology*. Chapman & Hall, London.
  Good in terms of the role of roads facilitating effects on destinations such as protected areas.
Southerland, M. T. 1995. Conserving biological diversity in highway development projects. *Environmental Professional*, 17: 226-242.
  Excellent overview with reference to the concept of bological diversity, the National Environmental Policy (USA), effects of roads at different stages with emphasis on fragmentation, environmental assessment of the effects, and mitigation.

---

from storm runoff to pollution control and habitat loss to heavy metals in roadside plants.

Some organisations, such as English Nature (UK), the World Bank, the Queensland Dept. of Main Roads (Australia) and the Federal Highway Administration (USA) have commissioned or prepared reports on the impacts of roads on nature conservation, which include guidance on mitigation (see Chap. 7). In the Netherlands, government-sponsored promotional publications on ecological effects of roads and well-publicised methods for dealing with habitat fragmentation are available (Chap. 4). The Land Use Planning Group of Wageningen University mainly focuses on the ecological effects of minor rural roads and the development of road networks (Dana Kamphorst and Rinus Jaarsma, pers. comm.).

Wildlands CPR in the USA is an organisation dedicated to addressing the effects of roads on nature. Of particular concern to the organisation are road verges in forested landscapes (Walder, pers. comm.). Wildlands CPR is a national network of grassroots groups and individuals working to reverse the adverse effects of roads. Their vision includes the following (Walder, 1998):

- Act as a national resource centre, providing tools to close or prevent environmentally damaging roads and motorised recreation in wildland ecosystems.
- Train activists to prevent, close and revegetate wildland roads, using sound biological and legal information to influence public land management processes.
- Educate the public about the environmental damage caused by roads and motorised recreation and how they can become involved in public land management decisions.

Publications of Wildlands CPR include *A Road-Ripper's Handbook* and a bimonthly magazine *Road-RIPorter* (Fig. 2.1).

Information on organisations interested in the ecology of roads and traffic is becoming quite common on the World Wide Web. Examples of such Web pages are listed below (addresses correct as of 2000).

Wildlands CPR
http://www.wildlandscpr.org

Critter Crossings
http://www.fhwa.dot.gov////////environment/wildlifecrossings/main.htm
http://www.tfhrc.gov////////pubrds/marapr00/critters.htm

University of Florida Transportation Research Center
http://uftrc.ce.ufl.edu/

Highway Safety Information System
http://www.bts.gov.NTL/DOCS/hsis/94-156.htm

NRVMA's Resource Room (deals with roadside vegetation management)
http://www.nrvma.org/resources.html

Laurance (2000): Roads in the Amazon and forest birds
http://www.ecoman.une.edu.au/postgrads/Susan-Laurance.htm

Lintermans, 1997: Review of wildlife reflectors
http://www.act.gov.au/environ/KAC3/KAC3-APD.html

Lintermans & Cunningham: Road-kills of kangaroos
http://www.act.gov.au/environ/KAC3/KAC3-APE.html

Some countries have held conferences on the ecological or environmental effects of transport. For example, an International Conference on Wildlife Ecology and Transportation (ICOWET) (now known as ICOET or the International Conference on Ecology and Transportation) has been organised four times in the last few years (*see* Evink et al., 1996, 1998). The most recent conference of this series took place at the Keystone Resort and Conference Center, Keystone, colorods (USA) in September 2001 (http://www.itre.ncsu.edu/cte/ICOET2001.html). The proceedings of these conferences include some very useful papers, in particular those addressing adverse ecological

The Road-RIPorter

Bimonthly Newsletter of the Wildlands Center for Preventing Roads. July/August 2000. Volume 5 # 4

Reclaiming the Concept of Restoration
A Story from the Pacific Northwest

By Jasmine Minbashian

Throughout the last century, thousands of organizations and individuals have been working to protect ancient and native forests, stop corporate subsidies on public lands, and protect our rivers and streams. The combination of their vigilance and a century of over-cutting have resulted in substantial declines in logging levels and stronger protections on our national forests.

In an effort to maintain timber harvest levels and generate public support for logging, land managers have employed "restoration" treatments to public lands. But these techniques, although sometimes creating the appearance of a mature forest, fail to protect ecological processes and therefore, the life that depends upon them. Jim Coefield photo.

The work, however, is far from over. Some would say it has only just begun. The aftermath of nearly a century of logging, grazing, and mining has left today's generation with a difficult legacy: more than 400,000 miles of logging roads, millions of acres of clear-cuts, and a drastic decline in native biodiversity.

— *continued on page 4* —

Fig. 2.1.

The Road-RIPorter

Bimonthly Newsletter of the Wildlands Center for Preventing Roads. March/April 2001. Volume 6 ǂ 2

Drive-Thru Wilderness?

Park Service Complacent With Cumberland Island Roads

— George Nickas

"...there shall be no commercial enterprise and no permanent road within any wilderness area."
— *Wilderness Act of 1964.*

Cumberland Island National Seashore is the largest undeveloped barrier island on the eastern seaboard. It is a surprisingly remote and biologically rich wilderness with the greatest diversity of biotic communities of any of Georgia's coastal lands. The eastern shore is marked by a glorious beach of white sand stretching the entire 17-mile length of the Island. Inland, the beach gives way to several rows of sand dune-communities, and then uplands filled with saw palmetto and a forest of pines and massive live oaks, draped in Spanish moss. The forest ends at the western shore, a rich mosaic of salt marshes and tidal creeks that provide a highly productive estuarine nursery. Over 300 bird species use the Seashore at various times of the year including the federally listed piping plover and wood stork. Over 50 species of herptofauna are present. American alligators are common, and the nesting population of threatened loggerhead sea turtles on the Island is one of the largest along the Georgia coast.

The Island narrowly escaped the large-scale development and devastation that besieged so many barrier islands when the federal government acquired most of the land and Congress designated it as the Cumberland Island National Seashore in 1972. In 1982, Congress went one step further and designated

Despite the long list of accolades and superlatives used to describe Cumberland Island National Seashore, the National Park Service seems content to allow its piecemeal degradation. Dr. Jerome Walker photo.

— *continued on page 6* —

Fig. 2.1. Two examples of *Road-RIPorter*. Publication of the Wildlands Centre for Preventing Roads.

effects. In 1997, the International Conference on Habitat Fragmentation, Infrastructure and the Role of Ecological Engineering was held in the Hague, the Netherlands.

The aforesaid conferences, together with the research, organisations and databases, are evidence that the ecological effects of roads and traffic are of concern to a wide range of people in different countries.

GEOLOGY, EROSION AND LANDSLIDES

Concerns about the effects of roads on the natural environment include the implications for geological conservation (Larwood and Markham, 1995).

Road construction in some localities presents paramount engineering challenges. Such challenges arise primarily from the geology of the region, inspiring such expressions as 'taming the geology'. Thus in the early 1990s, Puerto Rican officials constructed a road through some of the oldest and most scenic rain forest. Landslides, slumps, drainage problems and unstable alluvial materials and sinkholes presented many challenges, i.e., required 'taming the geology'. Celebration of the engineering feats overshadowed the ecological effects on the region—effects that will have implications on the wildlife of the area for evermore.

Road construction and roads of all kinds are likely to contribute to erosion and sediment deposition. During construction excessive erosion and sediment loads may enter watercourses; indeed, highway construction is often said to be the major source of sediment in streams. During winter freeze-thaw cycles and snow-melt may compound the problems of erosion and sediment production. Sealed roads increase the area of impervious surfaces and in doing so increase and concentrate water runoff. This runoff may contribute to erosion on road shoulders, in culverts and along embankments at the edges of bridges. Unsealed rural roads generally have less stable surfaces, which are susceptible to accelerated erosion. In the USA, roads are the main source of erosion into national forest streams (Criley, pers. comm.). Material from a road may also enhance sediment in any watercourse running between the road and the watershed. Management of water runoff and sediments has been the focus of many road-engineering reports.

Remedial measures have been widely discussed in the literature (e.g. Amaranthus et al., 1985; Fookes et al., 1985). Common measures include careful selection of the location and design of the road (avoiding steep slopes and highly faulted and sheared areas), design of road shoulders, grading of embankments, planting and landscaping slopes above and below roads, and design of roadside gulley pots and drainage systems.

Erosion from logging operations and forest roads can result in acceleration of surface erosion and mass movements, leading to severe erosion and sediment production. Indeed, it is widely agreed that road construction is a more important factor than deforestation in erosion acceleration. This is because in addition to the potential slope instability caused by roads (and deforestation), roads affect the patterns of water movement above and below ground, and due to the consequent changes in slopes above and below roads there may be mass movement of soil and rock material.

The debris slides and widespread erosion often associated with forestry have caused much concern, in particular their effects on nature. Loss of nutrients and soil has implications for productivity of the forest and sediments entering streams may affect aquatic species. The problems are compounded in some countries where forest roads facilitate illegal tree felling. In the Himalayas, tree theft is a major cause of forest loss (Haigh et al., 1995).

The type of logging operation; design, location and management of forest roads; and design and extent of buffer zones between the road and the forest affect the extent of erosion.

Water runoff, sediment deposition from roads and tree felling are major problems wherever roads are established in hills and mountains. The combined effects of roads and logging in montane areas can have major impacts on the environment, such as loss of vegetation, increased sediment load in watershed and landslides. In some regions, such as the Himalayas, road building is one of the major human impacts on slope stability; even the most carefully constructed roads may suffer massive disruption due to landslides (Haigh et al., 1993). Studies of landslide hazards along new highways or proposed highway routes have long been fundamental to road management and road engineering.

Road engineering and management of impacts in terms of landslides, water and sediment production are beyond the scope of this book. However, two broad areas have implications for ecology: 1) using nature to help mitigate the impacts and 2) the effects of changes in water and sediment production on organisms and biological communities.

LANDSCAPING

Besides engineering and safety considerations, landscaping has been a major contributor to road and roadside design and planting. Landscaping, plantings and types of trees or other plants used, can contribute to awareness of dangerous junctions and other potential hazards.

Of far greater concern than the ecology of roads (judging by the volume of literature) has been landscaping for aesthetic reasons and the visual enjoyment of motorists. Thus trees selected for visual appeal have been planted along road edges. For some, dead trees have added interest to the passing scenery. Others have considered dead trees untidy. Untidy or unkept roadsides have fueled passionate outcries for improvement of roadside appearances. So not only management of roadsides must be considered, but also the kinds of plantings. There has been much debate about roadsides serving as conduits for the spread of exotic and invasive plant species.

Although some may advocate the use of indigenous plants, the controversy of native versus exotic plants continues.

Sylvia Crowe in her 1960 book *The Landscape of Roads* wrote:

> While the basis of good road design lies in its siting, engineering and alignment, the right planting can both complete the efficiency of the road and a traffic-way and be the final link between road and landscape....Contributing to road safety, helping motorists to read the road, colours and textures, restoring the peaceful continuity of the scene—these have been the main criteria for landscaping roads.

Although much has been written about landscaping of roadsides, little attention has been paid to conservation interests. As a matter of fact, little attention has been paid overall to ecological and biogeographical opportunities. Opportunities to restore native plants, leave alone biological communities, have been present but often thwarted by landscaping with plants that meet perceived aesthetic needs.

In the UK, the 1990s seemed to herald a new interest in wild flowers. Wells and Bayfield's handbook (1990) *Wildflower Swards for Trunk Roads and Motorway Landscaping* prompted the notion that roadside verges could be enhanced with them for both aesthetic and conservation reasons. But in New Zealand, the trend is reverting to indigenous plant species for roadside landscaping (Fig. 2.2).

Fig. 2.2. Example of the use of indigenous vegetation on roadsides in New Zealand.

LIFE STAGES OF A ROAD

Road construction involves a number of stages: planning, building, making operational, managing, improving and, sometimes, decommissioning. The effects on the environment and nature differ at each stage.

During construction effects arise from the extraction, production and transport of materials for roads, including

— quarrying and transport of aggregates,
— production and transport of materials for road surfaces and edges,
— production and installation of signs, barriers, lights, tollgates, etc.

Many sites previously used for quarrying aggregates for roads are presently used as landfill sites for wastes. Roads have to be constructed to gain access to these landfills.

Filling such holes is a short-sighted measure and can lead to many environmental problems, such as seepage of chemical effluents. Vast amounts of methane gas are also produced. Developing methods to contain chemical effluents has cost millions of dollars.

In upgrading roads from gravel to asphalt, the bitumen fumes pose health hazards. Burgaz et al. (1998) report that bitumen fumes during sealing of roads can be absorbed by humans and polycyclic aromatic hydrocarbons (PAHs) may cause cytogenetic damage. These health hazards are being addressed and concomitantly research into pavement rehabilitation processes is underway. One new technology is the hot-in-place-asphalt process (Tideman et al., 1996), designed to meet the objectives of the Australian National Ecological Sustainable Development Strategy.

Roads as functional conduits for traffic require management of both the road and traffic. In turn, traffic requires fuel and a fuel industry, an infrastructure to manage traffic flow and an infrastructure to manage road traffic behaviour. Road management includes road washing and sweeping, application of de-icing agents, drain flushing and cleaning, and management of roadside shoulders. Mowing of grass verges has an effect on the verge wildlife community. Machinery used for mowing may transport seeds and weeds.

Decommissioned roads: Some roads become obsolete for one reason or another or engender extremely adverse effects. There are instances of roads (and railway lines) having been decommissioned (no longer used by vehicles) being managed subsequently as nature reserves or allowed to 'return to nature'.

ROADS AND PEOPLE

Aside from their main function—transportation—roads and traffic may affect humans in other ways, viz. effects on social behaviour. From road mortalities to roads bisecting villages and thereby influencing social behaviour, the general effects of roads and traffic on humans are the same as their effects on nature.

People are an inherent part of traffic; therefore, the environmental effects of roads and traffic may arise directly from human beings, e.g. pollution from litter, fires caused by unextinguished cigarette stubs and trampling of plants around road lay-bys. Roads have facilitated theft or lopping of trees for firewood in isolated areas. Motorways have extended the reach of criminals and made theft of livestock, equipment, ammunition and weapons easier.

Some roads have been constructed to open up land to agriculture or for mineral mining. Some are built as access roads to far-flung or remote areas. Such developments have sometimes caused extensive devastation of natural wildlife communities and landscapes. Rare plants nearby roads have been collected to extinction and trees felled for firewood.

Some large-scale road projects have had widespread impacts on natural habitats and people, resulting in huge losses of natural resources and ongoing social problems. The World Bank and Brazilian Government initiative in Rondonia in southern Brazil illustrates this well. A so-called centrepiece for the $1.5 billion development was the BR-364 paved highway. The financial cost may seem high but the environmental and social costs were huge and continue to rise. Loss of forest was rapid and massive, reaching 1.4 million ha each year, with roads playing a decisive role in this loss (Reid and Bowles, 1997).

Road transport also engenders health, safety and social issues. In some cities, problems of traffic pollution are severe. In Kolkata, West Bengal, studies on traffic personnel have shown high concentrations of lead in blood tests done on those experiencing the greatest exposure to congested road intersections (Mukherjee et al., 1998). Patna in Bihar (north-east India) is one of the fastest expanding cities in the world, with pollution levels rising by 200% over the last four years (Trivedi et al., 1993).

Pollution caused by vehicular emissions (CO, CO_2, HC, NO_x, hydrocarbon, ozone, particulates and lead) and pollution from tyre wear contribute to the rise in environmental and health problems (Detwyler, 1971; Watkins, 1981; ECMT, 1990). Noise pollution exists everywhere, especially in urban areas. Deaths and injuries abound. Roads may displace populations and require resettlement of communities. Some communities may be disrupted and divided by roads; new roads may bring about more developments resulting in rapid urbanization and demographic changes. Traffic congestion contributes to transport costs.

ENVIRONMENTAL EFFECTS (EXCLUDING ECOLOGICAL)

Environmental effects of roads and traffic (Table 2.2) include energy and pollution issues. Transportation on roads is dependent on the use of fossil fuels. The use of resources such as metals may in the long term lead to recycling problems. Land is taken up for roads. Landscapes are changed and sometimes 'scarred' by roads. Pollution from roads and traffic may affect the

Table 2.2: Environmental effects (excluding ecological) of roads and vehicles
Note: A more detailed summary of ecological effects is given in Table 2.3

General comments
- Effects differ for the different stages of a road: construction, operation, maintenance and decommissioning.
- The scale of roads ranges from a dirt track to an 8-lane highway and from single roads to road networks.
- Urban developments are dominated by roads, car-parking facilities and infrastructure supporting vehicular use.
- Generalisations about the effects are difficult since the types, level of use and locations of roads differ.
- Spatial and temporal scales of effects exist. Some pollution may be localised while some contributes to global pollution.
- Spatial and temporal issues are important when determining the extent of the effects. For example, should the effects of dredging or quarrying for aggregates (for road surfaces) be included. This is one of the problems faced by authors of EIAs.
- Secondary and tertiary effects, cumulative effects and synergistic effects must also be taken into account.

A. Non-sustainable use of energy
Most vehicles on roads have internal combustion engines; use of petroleum products is dependent on a non-renewable resource.

B. Pollution
Global climate change: CO_2 from combustion is a major contributor to global climate change and smog in cities.

Microclimates: Roads create new microclimates, causing loss of individual organisms and species and changes in species composition.

Other pollution effects, mainly from fuel, engines and tyre wear (CO, CO_2, HC, NO_x, CFC, particulates and heavy metals). There is also pollution from transported materials (hazardous chemicals, effluent spillage).

Pollutants may affect air, soil and water quality. Some pollutants may contribute to erosion of buildings and statues. Nitrogen deposition in soils affects the ecology of soils and plants and heavy metals accumulate in animals.

Visual (fumes, smog and dust), amenity (fumes and odour) and health effects can be traced to road vehicles. When roads are upgraded and sealed with asphalt, health hazards arise from polycyclic aromatic hydrocarbons (PAHs). These can be absorbed through the skin and/or inhaled.

Solid and liquid waste materials: Waste material from road construction, waste oil, disposal of tyres and accumulation of vehicles withdrawn from use. Litter is deposited on roadside rest areas, picnic sites and lay-bys. Illegal dumping of wastes takes place.

C. Disturbance and perturbations
- Vehicular noise, traffic movement, vibrations, vehicular and road lights, and dust.

Contd.

Table 2.2: (Contd.)

- Vibrations may have implications for structural damage of old and historical buildings.
- Some physiological effects on humans and other animals are known. Road rage has become a common phenomenon. Stress resulting from driving is well documented.
- Some animals experience behavioral effects that disturb communication (primarily due to noise) and/or flight.
- Some vertebrates and invertebrates avoid areas near roads.
- Some animal/insect species are attracted by traffic lights and road lights.
- Traffic vibrations affect some animal/insect species.
- Wind gusts produced by vehicles may affect growth of some plant species.
- Dust deposits affect photosynthesis of some plant species.
- Fires caused by vehicles and litter disposed on roadsides. Roads and road networks contribute to incidence and patterns of roadside fires and subsequent patterns of regrowth in vegetation.

D. Effects of road maintenance, roadside management and landscaping

Herbicides: The general use of herbicides has implications for the local flora and fauna.

De-icing agents: Some de-icing agents alter plant species composition. Salt on roads attracts some animals.

E. Effects of road infrastructures and associated structures

- Parking facilities have effects on urban ecology.
- Bridges, tunnels, viaducts, causeways, median strips or central reservations have effects on fauna and flora.

F. Effects on the landscape and landscape ecology

Visual effects: Roads impact visually on landscapes. Infrastructures for roads (petrol stations, cafes) and associated structures (road signs) impact on vision.

- Landscaping and planting along roads are aesthetically pleasing.
- Roads contribute to cumulative effects on the environment.
- Roads provide access to undisturbed areas, leading to cumulative effects in land-use changes, introduction of resource use (legal or otherwise), recreation and urbanisation.
- Roads may affect the movement patterns of animals.
- Roads may be a barrier to wildlife movement.
- Linear habitats created on verges are colonised by new species.
- Roadsides are used by some species for feeding and basking.

Corridor effects: Dispersion of alien and pest species.

G. Geological effects of roads

- Effects increase according to the type of road, geology and geography.
- Increased impervious surfaces lead to changes in water runoff patterns and erosion and sediment production. Increased runoff into coastal areas may affect the marine ecosystem.

Contd.

Table 2.2: (Contd.)

- Erosion and landslips may cause loss of vegetation.
- Erosion may remove topsoil; the diminished quantities of nutrients affect productivity of forests and forestry.
- Sediment may enter watersheds and affect aquatic organisms. This may have implications for recreational and commercial fisheries.

H. Physical effects of vehicles

- Deaths, injuries and property damage.
- Large mammals crossing roads cause accidents.
- Road kills provide food for some species and attract scavengers.
- Vehicles act as agents for transport of organisms. Pest species can be dispersed.

I. Social effects

- Partitioning or destruction of neighbourhoods.
- Road bypasses have both positive and negative social and economic effects.
- Improved access to resources such as forest, which in turn may lead to tree theft.

J. Effects on land use

- Land used for temporary construction activities.
- Land used for the road and road verges.
- Land used for the stabilisation of roadside edges.
- Land used for road transportation infrastructures such as parking, extraction of road-building materials.
- Associated activities include vehicle sale yards and wrecker yards.
- Habitat loss and fragmentation.
- Individual organisms are displaced or destroyed.

K. Effects on geology and hydrology

- Roads cause soil compaction, erosion, surface water runoff, changes in groundwater movements and modification of water systems.
- Physical alterations in aquatic habitats may take place.
- Introduction of new materials (road aggregates and sealants) affects the chemistry of surrounding soils.
- Soil pH is altered, causing changes in plant composition.
- Effects on hydrology alter aquatic communities.
- Gulley pots trap amphibians.
- Some species use road grit.

L. Long-term effects

There is a slow but overall incremental effect, the implications of which humans find difficult to recognise because it develops over decades.

quality of 'health' of waterways, air and soil. Degradation and erosion of agricultural and forest soils may occur.

Environmental effects of roads may also be seen in structures and utilities associated with roads and traffic. For roads, these include bridges, tunnels, lights and lighting, signs, culverts, lay-bys, central reservations, cat's-eyes and road markings. Movement of traffic and humans through or on these structures has ecological effects. Litter may accumulate at lay-bys, birds use the underside of bridges for roosting and nesting, bats may use some tunnels and aquatic life may establish in culverts.

Structures and developments associated with traffic include car parks, garages, fuel stations, petrol pumps, vehicle sale yards and wrecker yards. In the UK, there is an 'environmental tax' for car wreckers. This cost is passed on to motorists for wrecking their car. The cost of car wrecking has thus increased and more and more cars are now being abandoned in the countryside. Many dumped cars are set on fire. This dumping and sometimes retrieval methods result in damage to hedgerows and grass verges.

The incremental and ramifying effects are the most important of all environmental effects. Roads lead to more roads (Box 1.4) and other developments. They are similar to aging. One is unaware of it when young and hardly notices it day by day until suddenly its effects announce themselves!

Construction of a single new road almost certainly increases the chances of more roads extending from that already built. Southerland (1995) has drawn attention to the incremental effects of road developments and argues in favour of roads as an important cause of loss of biological diversity. Incremental effects are by far the most damaging but perhaps the most difficult to appreciate in the lifespan of a human being. New road proposals will incite arguments against their being built. Inevitably, the road will nonetheless be constructed, be it on a modified route or in some modified form. Another generation goes by and roads extending from the original road are proposed. The old arguments against such are forgotten but new objections put forward. Sadly, these objections are not strengthened by the arguments of the previous generation. Meanwhile and additionally, roads have facilitated industrial and urban development and provided an infrastructure for tourism. The incremental ecological effects thus continue.

New roads in undeveloped areas quickly result in large-scale effects on natural habitats. They enable travel and opportunities to visit new and possibly undisturbed areas. Therefore new roads may impact on tourism in new areas. This aspect is not considered in the present study. However, it is important to note that studies have been done on human disturbance, including trampling and physical disturbance of richness of plant species in 'isolated' conservation areas. Drayton and Primack (1996) in their study of a 400-ha woodland park in Boston found an increase in exotic species and a decrease in indigenous, which correlated with an increase in human disturbance (see Chap. 4 for more details).

Roads and road networks have a definite affect on nature. Various road densities and their effects have been investigated (see Theil's work (1985) on road densities and wolf habitats, for example). The most important implication of road networks is habitat fragmentation, which is discussed in Chap. 4.

The magnitude and scale of environmental effects (Table 2.2) depend on geography, climate, current land use of the area and traffic density. Roads on flat land in temperate climatic countries may have in short- and medium-term, few noticeable ecological effects. Contrarily, the immediate and short-term ecological effects of roads in the Himalayas are readily discernible: massive destruction of natural forests often leading to landslides, and accelerated erosion. In general, roads affect 2.5-3.5 times the landscape than the surface area occupied by the road per se (Reed et al., 1996).

The incidental effects of roads include crime and exploitation of nature and natural resources (Box 2.1). Should we accept that these are just incidental or ought these effects to be included in cost-benefit analyses?

Box 2.1: Incidental Effects of Roads

Orchids and other plant species can be rescued from road-building crews. Eric Hansen (2000) describes in his book *Orchid Fever* how more than 15,000 orchids have been salvaged in Minnesota (USA).

Hunters in Cameroon have found the expanding networks of logging roads very convenient for hauling their kills of monkeys and apes to market (McRae, 1997).

Motorways out of metropolises have been a boon to organised burglary. Newspaper reports of such activities increased during the 1980s for southern England.

Roads in the Himalayas have facilitated illegal logging/lopping of trees for firewood (Haigh, pers. comm.).

ECOLOGICAL EFFECTS

Roads and traffic affect nature; conversely, nature can affect roads. For nature's effects on roads example in northern Ireland, badgers undermining roads is a common cause of road subsidence. In the countryside, roads tend to follow higher, drier ground and likewise badgers! So, wherever possible, reinforced concrete decking is used to support the road surface over badger dens.

Plants are sometimes used in landscaping roadsides and in inner city areas trees have proven helpful in blanketing noise, vehicular lights and even airborne pollutants. In other words, both animals and plants exert an effect on traffic and in certain circumstances constitute an integral part of the road environment.

It is sometimes said that because of the extent of roads and the many kilometres of roadside verges serving as nature reserves, roads benefit nature. Use of roadside edges for nature conservation is just a very small compensation for habitat loss resulting from roads and cannot be considered a net gain in terms of benefits.

The ecological effects of roads and vehicles are many and varied (Table 2.3 and Fig. 2.3) and have been well or minimally documented in texts on life

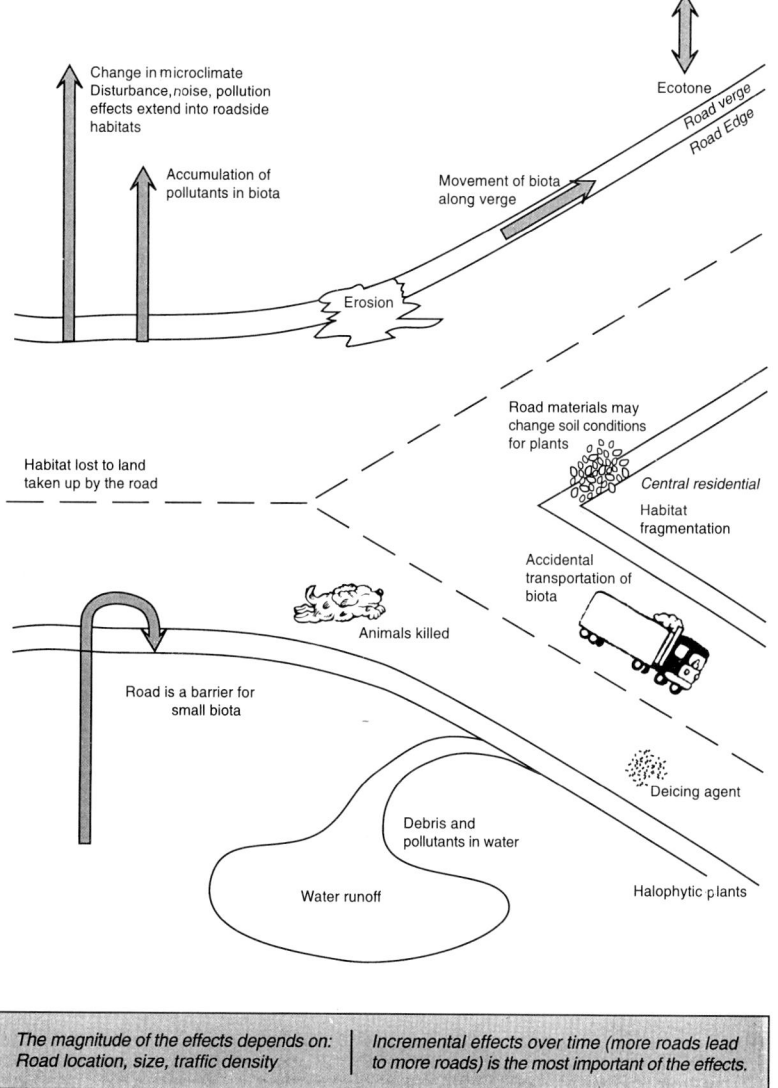

Fig. 2.3. Simplified pictorial representation of the ecological effects of roads and traffic (compare with Fig. 7.2, Chap. 7).

sciences and transport planning. O'Flaherty (1997), for example, notes that roads have many detrimental effects on ecology, notably:

— direct loss of habitat,
— road mortalities,
— effects on hydrology,
— effects of pollution,
— road lighting adversely affecting some invertebrates and birds,
— salt on roads attracting mammals.

36 Ecological Effects of Roads

Table 2.3: Main ecological effects of roads and traffic on nature

Undoubtedly, the greatest effect of roads and vehicles on nature is incremental and cumulative. Roads beget roads and concomitantly other developments.

The cumulative and progressive effects of roads and traffic on nature—when typically little happens over a long period of time—make recognition of the magnitude of these long-lasting and extensive effects difficult. But as already pointed out, one new road leads to more roads over time, which entail many new projects. The consequences for nature can be devastating and often irreparable.

A. Effects during construction

Effects caused by mining or manufacturing materials for roads. Land taken up for construction equipment, construction teams and temporary roads. During road construction per se:

- Loss of habitat (usually not quantified in terms of area) and biota.
- Soil compaction with a long-lasting effect.
- Erosion from new embankments and road cuts.
- Additional effects (add pollution to the above) due to supporting activities for construction, e.g. construction site and composite activities.
- Disposal of soil and rock from excavations can affect local soil chemistry and plant communities.
- Impacts occur beyond the immediate vicinity of the road, e.g. changes caused in the hydrology. Quarrying for aggregates for road making may be done in a different area. It is thus important to agree on the geographical boundary for an impact assessment.

B. Effects from upgrading

When gravel roads are upgraded by sealing with asphalt or bitumen, secondary ecological effects occur. Sealed or paved roads continue to be a source of chemical pollutants long after upgrading (Criley, 2000). Typically the road is widened and this increases the habitat fragmentation effects. Vehicular speed increases, augmenting the risk of animal mortality on roads.

C. Effects when roads are operational

i) *Short-term effects*

— If vegetation is removed, the new gap (linear gap) creates a new microclimate and a change in physical conditions extends varying distances from the road edge.
— Soil compaction takes place.
— Upon removal of vegetation, plant mortality will occur in the short term and extend varying distances from the road edge.
— Plant mortality exerts both a direct and secondary influence on other organisms.
— Some fauna will move from the area of the road as a result of habitat loss and physical disturbance.
— Animal road kills are likely.

Contd.

Table 2.3: (Contd.)

ii) *Long-term effects*
— Road kills have secondary effects as a source of food. They attract scavengers. The distribution of road kills may alter the distribution of birds of prey.
— Loss of habitat and change in habitat extend beyond the edge of the road. Ecological effects may extend some 200 m into forests.
— The land take necessary for roads may be at least twice the area of the actual sealed road.
— A change occurs in biological communities, ranging varying distances from the road edge, sometimes up to 200 m.
— Habitat fragmentation occurs which in turn has implications for habitat damage and loss, dispersal and vagility of organisms and isolation of populations.
— The newly created edge provides habitat for edge species.
— The edge habitat or ecotone and road traffic may facilitate dispersal for some taxa, resulting in dispersal and establishment of alien and invasive species.
— Dispersal of pest species via ecotones or traffic may have secondary effects on biological communities.
— Associated structures such as bridges and tunnels may provide habitats for some taxa.
— Aggregate and sealant materials used for road construction may affect the chemistry of surrounding soils and subsequently the plant communities, e.g. calcareous road materials on acid soils.
— De-icing salts and other chemicals affect roadside plants and support halophytic communities.
— Impervious surfaces of roads and the presence of culverts alter the local hydrology. Water in culverts may alter water tables.
— Surface water runoff from the roads may contain pollutants and affect aquatic communities.
— Exhaust emissions, pollutants from tyre wear, dripping oil, litter, noise and other physical disturbances may extend into roadside vegetation for varying distances, leading to changes in species composition.
— Traffic noise may affect communication between bird species.
— Dust caused by traffic, especially on unsealed roads, may affect photosynthesis (Farmer, 1993).
— Wind gusts from passing traffic may affect plant growth.
— Roads may act as barriers for some taxa, limiting dispersal and vagility. Secondary effects may develop due to changes in dispersal of some predators.
— Roads provide an infrastructure for more developments and incremental effects. One road encourages construction of another.
— Roads bring people and other impacts into undisturbed areas.
— Roads and road networks influence where fires start and the pattern of their spread as well as subsequent regrowth of vegetation.

iii) *Secondary effects of roads and traffic on nature resulting from:*
— Hydrological changes consequent to increased impervious surfaces.
— Introduction of new chemicals from aggregates.

Contd.

Table 2.3: (Contd.)

— Unstable hillside slopes with sediment build-up in waterways.
— Litter deposited on roadsides, picnic areas and lay-bys.

D. Features and activities associated with roads which may have ecological implications

During construction

— Construction sites.
— Temporary housing and office sites.
— Quarries for extraction of hardcore material for road building.

Following construction

— Car parks and lay-bys
— Bridges
— Tunnels
— Culverts, ditches
— Divider strips and roundabouts
— Road signs
— Road lighting
— Wrecker yards
— Use of old quarry sites for landfill—introducing a new set of ecological issues.

SHORT- AND LONG-TERM EFFECTS

Effects on roads and traffic can be divided into 1) short-term, resulting from the construction of a new road and 2) long-term, once the road is operational. A striking example of the most immediate effect of road on nature is that of one drilled through the base of a giant redwood tree in Sequoia National Park (California, USA) (Fig. 2.4). Was this a statement about human dominance over nature?

Short-term effects cause changes in the microclimate and destruction of habitats in the wake of road construction activities. Microclimatic changes exert an effect on individual organisms, thereby changing the species composition and structure of the roadside biotic community. Long-term and cumulative effects may be the result of pollutants entering the ecosystem, the result of fragmenting a natural biological community or inevitable effects of developments arising from the transportation benefits provided by the road.

DIRECT AND INDIRECT EFFECTS

Direct effects of roads on nature and the environment are determined by type and management of the road, traffic density, physical geography, geology, weather and climate. The effects include incremental ones over time, loss and fragmentation of natural habitats, isolation of habitats and creation of new habitats on road edges and verges. Linear edge habitats may act as wildlife

Fig. 2.4. Road through the base of a giant redwood tree (photo, Charlie Challenger).

corridors and facilitate movement of wildlife across the countryside and from one habitat to another.

Crossing a road may be risky for a small mammal, amphibian or reptile; for other animals, the open road space may act as a deterrent to crossing over. Roads may thus serve as barriers for some wildlife, precluding or limiting movement of animals from one side to the other. Sometimes fences are erected along roads to prevent large mammals from crossing. A road barrier affects dispersal of some species and may even hinder the necessary migration of others.

Once roads are established, the physical linear feature, traffic with its noise and lights, and the social behaviour of people resting in lay-bys may attract some animals to roadside edges, lay-bys and picnic spots.

Some forms of wildlife benefit from roads and the structures associated with them. Roadside edges provide basking areas for reptiles and the dust and gravel of unsealed roads provide birds dusting sites and sources of grit. Road carrion constitutes a food item for some animals.

Road edges provide habitats for some foms of wildlife. Road edges and ditches can support many plant species such as orchids. As noted earlier, tunnels, bridges and other structures associated with roads are used as roosting sites for birds and bats.

Perhaps the most obvious and well-known effect of traffic on wildlife are animal fatalaties (road kills). Road kills directly affect wildlife and is a very significant cause of mortality in species such as armadillos. An indirect effect is the attraction the carcasses hold for scavengers, in particular some birds of prey, notably buzzards, kestrels, harrier hawks and seagulls.

Roads are used for the deliberate transportation of plants, plant material, animals, livestock and animal materials. Some environmental effects are directly or indirectly attributable to transportation of such organisms. Outbreaks and spread of foot-and-mouth disease have been augmented by livestock transportation. Other impacts include spillage of livestock effluent and runoff from stockyards.

Weed seeds may be transported and dispersed by vehicles. The ever-increasing road networks with rapid transport systems may contribute to the spread of diseases; this may have led to the outbreaks of foot-and-mouth disease in the UK in 2001.

Traffic has other effects as well. Disturbance and displacement of animals can result from traffic noise, artificial lights and traffic movement. Wind gusts caused by fast-moving traffic may affect some plants and animals.

Water runoff from impervious and compacted surfaces into natural water systems can affect aquatic biota.

Pollutants come from vehicles, people in vehicles and road surfaces. Plants and animals can be affected by pollution.

Direct effects of roads and traffic on nature may result from pollution from engines, tyres and noise. Chemical effects of aggregates used for roads (e.g. base materials may have an impact on acidic soils), dust pollution (unsealed roads) and materials from sealed roads may also affect wildlife. Where roads have been constructed through hills or around mountains, there will almost certainly be erosion. Erosion from roads on steep slopes may result in debris entering watercourses and thereby affecting aquatic species.

There are also many indirect effects from both roads and traffic as well as the structures and utilities associated with them. Where cuttings and embankments have to be established, revegetation of bare slopes brings with it a new wildlife community structure, which in turn affects the biogeography of the region.

De-icing of roads results in accumulation of chemicals, including salts, which in turn increases the abundance of salt-loving or salt-tolerant plants over non-tolerant species. Similarly, through the use of waste oil on unsealed roads for dust management, oil may contaminate water bodies, which may have toxic effects on aquatic wildlife.

HABITAT FRAGMENTATION AND BARRIER EFFECTS

Roads contribute to habitat fragmentation. This has occurred on a massive scale in the Amazon in particular, as well as in other parts of the world. Habitat fragmentation is bad news for many animals. Habitats are destroyed or greatly reduced in area. Natural migration routes are disrupted. Roads constitute barriers for many small flightless animals. Smaller habitats may result in fewer resources for some wildlife.

The area taken up by roads results in habitat loss—and the effect does not end at the road edge. Pollution, changes in microclimate and general disturbance may extend to 30 m, sometimes even 100 m, from the road edge. Reed et al. (1996) have estimated that roads affect 2.5-3.5 times the landscape than the area of land taken up by the road. Recent research shows that some effects resulting from physical disturbance, microclimate changes and soil compaction may extend some hundreds of metres from the road. What is becoming known as the 'road effect zone' (Forman, 2000) is much more extensive than previous studies have suggested. So the effects of roads are not limited just to the narrow road shoulders.

WHAT IS TO BE DONE?

How can the environmental effects of roads be reduced? Several means have been proposed, ranging from local incentives to formal planning procedures and Environmental Impact Assessments (EIAs). The following is a brief overview; more detailed accounts are given at the end of Chaps. 4, 5 and 6 and elaborated in Chap. 7.

Location and design of roads and design and extent of road verges play an essential role in reducing environmental effects. More envionmental information is needed when identifying options for new road routes. This includes data on location of heritage sites, protected areas, sensitive habitats, including wetlands, movements of large mammals, populations of threatened species and areas of conservation importance.

Planning procedures and EIAs are essential in reducing environmental effects. An EIA is a formalised process in which the impacts (effects) of projects are identified and assessed (see Chap. 7 for details). This formal process is mandatory in many countries. EIAs include social assessment and ecological assessment. An Ecological Impact Assessment (EcIA) is a formalised process in which the effects of a project on the ecology are determined.

So, EIAs or EcIAs help in identifying unwanted effects, which is not to say that anything will be done about them. Furthermore, some legislation refers to 'significant' effects but fails to define the term 'significant'.

Deciding where the effects end or how far they extend is an issue common to EIAs and EcIAs. For example, spatial limits to the effects of a road might be considered to end at the physical boundary of the roadside. However, pollutants from the road surface and traffic may be carried some distance from the road. Similarly, construction of a road requires transportation of aggregates. Should the EIA therefore cover the source of the aggregate materials? It is difficult to say. The important thing is agreement on the spatial boundaries. Many effects can be avoided and reduced prior to road construction and prior to selection of its route. As a matter of fact, assessing the likely impacts of a road and its traffic on human health and safety as well as the likely effects on the environment has been a standard procedure in road construction and development for a long time.

Management processes are now readily available that help avoid, reduce and mitigate environmental effects. Unfortunately, the range and extent of the impacts of roads and traffic on nature do not seem well understood and the practice of EIAs and EcIAs has been undertaken in a less than satisfactory manner.

Long before the formalised and structured concept of EIAs arrived on the scene, road project managers were undertaking assessments and cost-benefit analyses of roads and traffic. EIAs came to the fore in the USA in the 1960s and since then have been made part of the statutory environmental education in many countries. However, the use of EcIAs does not have a good record and the standards of many ecological impact assessments have been poor.

Perhaps the most commonly known aspect of the ecology of roads is wildlife kills. Local incentives to protect wildlife near roads have even taken the form of awards. For example, in New Zealand, the Birdlands Sanctuary Trust (Christchurch) annually seeks nominations for people, clubs, societies or firms working for the environment through their own projects. In 1999, the Business Award went to the construction company Fulton Hogan for removing road kills each day from a highway on Banks Peninsula, which resulted in fewer hawks being struck by cars.

Roads take up a great deal of land and so diminish the area for wildlife habitats. With appropriate management, some of that loss can be offset by establishing native wildlife plantings on road verges. Grass verges on roadsides are frequently mown to within a few millimetres in height. By simply changing the mowing regimen, the diversity of plant and invertebrate wildlife can be greatly increased. Where safety considerations allow, grass should not be mown, which sometimes results in a beautiful grass sward.

Loss of habitats is a serious issue as is the way in which roads fragment habitats. An innovative, bold and intriguing effort to reduce fragmentation effects of new roads and traffic on wildlife habitats can be found in some areas in Europe. Driving on motorways in Europe may seem like driving on any motorway or freeway anywhere in the world. But there is something very odd about some of the bridges that cross the motorways here. Some of these bridges are covered with soil, grass and shrubs; such bridges are not for traffic but rather for nature!

What does it take to persuade someone to spend money on bridges for nature? Who pays for these bridges? Where should such bridges be positioned? Bridges for nature are called ecoducts or green bridges (see Chap. 4 for details and Fig. 2.5 below). They were built in an attempt to solve the issue of roads as barriers to wildlife movement and the issue of road fragmentation of habitats. They symbolise a growing recognition that the effects of roads and traffic on nature have been serious—serious enough to warrant searches for solutions. However, some interesting biogeographical issues were not fully addressed during the planning of these bridges. In brief, are they located at the best possible sites and do they work?

Fig. 2.5. An example of an ecoducts in the Netherlands. This is the A50 near Woeste Hoeve. From 'Nature across motorways' (1995, Rijkswaterstaat (RWS), Dienst Weg- en Waterbouwkunde (DWW), Deft) with permission of Hans Bekker).

When thinking about the ecological effects of roads, one should not forget that the urban environment is dominated by both buildings and roads. The ecology of urban roads has not been well studied yet attempts have been made to address the problems of traffic density in urban areas. One way to solve this problem and the effects of vehicular pollution is road closure or conversion of streets into pedestrian precincts. Motorised traffic has now been banned from some inner city streets and bridges, in an effort to preserve pedestrian havens and restore what once were thriving retail outlets.

For many decades much was written about the design of roads and management of traffic (see articles in such journals as *Transportation and Research*). Transport study has become a discipline of its own, building largely on engineering and landscape studies. The last decade has seen a notable change. The effects of roads and traffic have begun to expand beyond human health and safety to incorporate plants—and not just those used in road design—and animals—their habitats and biological communities. Road policy is starting to include ecology. In some countries, notably Australia, Norway, the Netherlands and the UK, detailed management and best-practice manuals have been published, which provide answers as to how adverse ecological effects can be reduced (see Chap. 7).

ROADS AND TRAFFIC IN PERSPECTIVE COMPARED TO OTHER ENVIRONMENTAL EFFECTS

It is difficult to estimate the magnitude of ecological effects of roads vis-a-vis other impacts—agricultural, forestry, mining and urbanisation. But I feel they should be accorded the same level of importance as climatic changes. The area of land occupied by roads, the long-term ramifying cumulative effects of both roads and traffic on the environment, and the losses to nature are immeasurable and will be with us for many years to come. However, there are signs of a turning point. In 2000, the then President, William Clinton, established or expanded ten national monuments in the USA and decreed that no roads would be built in the 23.5 million hectares of inventoried roadless areas within National Forest Lands. It is hoped that the new President of 2001 will not reverse these decisions.

3

Biology of Roads and Roadside Verges

INTRODUCTION

Despite long interest in the relationship between roads and nature, the biology of road right-of-ways, roadside verges, central reservations and junction reservations does not seem to have attracted much attention. Only a handful of systematic inventories on the species encountered (mostly, the birds) are available. The paucity of studies appears very strange for two reasons. First of all, the area of roadside verges and central reservations is very extensive in some countries. Surely something so extensive and visible as roadside verges ought to have attracted more interest. Secondly, it has often been said that roadside verges could make a contribution to nature conservation, especially if managed in an appropriate manner. That being the case, expectedly more interest ought to have been shown in the kinds of wildlife that could benefit from roadside nature reserves.

So, why so few studies on the wildlife of roadside verges? Is it because we pass them by (don't enter them) that they remain unnoticed. All road verges are man-made or at least modified; in urban and suburban areas in particular, trees, shrubs, grasses and herbs have been planted and the habitats are thus entirely artificial. Therefore lack of interest might be due to some preconceived idea that habitats so modified, so disturbed and so polluted would hold nothing of interest. Or does the risk of accidents while undertaking surveys of roadside verges pose a threat?

This chapter is mainly concerned with the question—what biota is associated with road right-of-ways and roadside verges (see Appendix I for definitions). Information on the biology of median strips and islands at road intersections is also included. Are there any plants or animals characteristic of road verges? Interest in the wildlife of roadside verges centres today in live plants and animals. Dead trees alongside roads were once considered noteworthy for their contribution to natural habitats if not landscapes.

In some countries, roadside verges are the last remaining vestiges of wildlife habitats. The potential of road verges as nature reserves is not a new idea and the contribution they could make to conservation is discussed. Lastly, if road verges can contribute to conservation, then some ecological restoration would be appropriate. How are road verges being managed for conservation and are the methods employed the best? Any number of questions could be asked but one thing is certain: costs of management of roadside verges can be reduced by incorporating wildflower swards and by less mowing.

TRAFFIC AND DISPERSAL OF SPECIES

For thousands of years mankind has contributed to changes in the dispersal of plants, animals and other organisms. We have mixed and stirred the world's biogeography either deliberately or by accident. Some people speak of homogenisation of the world's biota. Many species have been deliberately introduced in new areas, either as a new source of food (spread of cereals around the world), or as an agent for pest control (introduction of the moth *Cactoblastis cactorum* to eradicate the prickly pear cactus) or as ornamental plants.

Given the increase in volume of road traffic, it is no surprise that it has aided in the distribution of species (see Chap. 4). Road vehicles (as well as trains, barges, ships and aircraft) have all contributed to species dispersal and consequently affected the biogeography and species composition of many areas of land and sea. The infrastructure for transport has likewise aided species dispersal. Roadsides (as well as edges of railway tracks and banks of canals) have all acted in some way as linear habitats and facilitated the dispersal of many species, while concomitantly hindering that of others.

SURVEYS AND INVENTORIES

When studying biology in the field (in contrast to studies in botanical or zoological gardens), some important and first questions must be asked, namely: 'what is there'?, 'what's the abundance'?, 'what's the distribution'? and 'what condition are the organisms in'? Such questions generally refer to the biota of a given area but are also relevant for verges of established roads. What organisms, habitats and communities are found alongside roads, why are they found there or what determines their presence in that particular site?

For roads recently established, interest is more likely to centre on the effects of the new road per se rather than the wildlife remaining on or near the verge. Contrarily, in urban areas and where established motorways cut across industrial and agricultural landscapes, interest is more likely to focus on what inhabits the roadside verge rather than the impact made by the road.

But why undertake an ecological survey of a roadside verge, aside from ascertaining what is there? Because knowing what is there helps identify species or communities not well represented elsewhere. Endangered (in the sense of listed in a red data book such as the *UK Red Data Book*; see Perring and Farrel, 1977) habitats or species may be present or some protected species (in the sense of statutory protection).

Basically, knowing what is there provides a basis for studying the ecology of organisms and that in turn is a necessary basis for management of the roadside verge.

Relatively few results of general surveys of roadside verges have been published (in English anyway), enabling development of inventories or databases (Table 3.1). Some surveys were undertaken as part of nation-wide surveys to identify natural areas for protection. Other surveys were undertaken as part of the research on the effects of roads and traffic (e.g. the effects of

Table 3.1: Examples of general biological surveys and inventories of wildlife in roadside verges reported in the literature

Hawaii
Wester, L. and Juvik, J.O. 1983. Roadside plant communities on Mauna Loa, Hawaii. *J. Biogeography*, 10: 307-316.

New Zealand
Given, D.R. 1994. Roadsides, railway margins and waterways. Forgotten natural habitats. NCCB.
Ullman, I., Bannister, P. and Wilson, J.B. 1995. The vegetation of roadside verges with respect to environmental gradients in southern New Zealand. *J. Vegetation Science*, 6: 131-142.
Wilson, J.B. et al. 1992. Distribution and climatic correlations of some exotic species along roadsides in South Island, New Zealand. *J. Biogeography*, 14: 183-194.

The Netherlands
Haeck, J., Hengeveld, R. and Turin, H. 1980. Colonization of road verges in three Dutch polders by plants and ground beetles (Coleoptera: Carabidae). *Entomologia Generalis*, 6: 202-215.
Schaffers, A. 2000. *Ecology of roadside plant communities*. Ph.D. thesis, Dept. Environmental Sciences, Wageningen University, The Netherlands.

Vienna.
Holzner, W., Kriechbaum, M., Kutzenberger, H. and Bohmer, K. 1989. A good summary of the importance of roadside verges for nature conservation and they also relate their work to similar work in Germany and The Netherlands.
Thaler, F., Bohmer, K., Kriechbaum, M. and Holzner, W. 1996. A deailed survey of some Austrian roadside verges.

UK
Cheshire Ecological Services. 1995. Cheshire roadside verge survey.
Reeve, H.A. 1977. Evaluation of roadside verges. *Watsonia* 11: 148-149.
Feltwell and Phillip (1980). This is about the natural history of the M40 Motorway.
Way (1976). This is about vegetation surveys of motorways in the U.K.

USA
Harper-Lore, B. and Wilson, M. Eds. (2000). *Roadside use of native plants*. Washington, Island Press.
Nilson, D. 1977. Roadside management and wetland development along North Dakota Highways. *North Dakota Outdoors*, 40: 23-25.

pollutants from roads on biota). Such surveys provided the data for species inventories. Most are concerned with birds and plants and very few with small mammals, reptiles and invertebrates. Surveys on the latter are particularly scant.

HABITATS AND PLANT COMMUNITIES

Surveys leading to species inventories are a first and important step towards studying the biology of roads and roadside verges. An inventory (of any taxon) is useful and a very basic piece of information. Similarly, surveying and identifying various roadside habitats is a very useful exercise. Information on habitats provides a simple basis for describing any variation in conditions. However, there appears to be no formally recognised classification of roadside habitats and therefore naming them is purely arbitrary. Such could be based on the dominant features or the dominant type of vegetation. For example, there are ditch habitats, grass verge habitats, shrub habitats and hedge habitats (examples of habitat types identified in the Cheshire Road Survey (U.K) are shown in Box 3.1 presented later). Aside from habitats, roadside verges could be divided into linear zones (see the RSNC leaflet; Box 3.2) from those at the very edge of a paved road to those farthest from it and adjacent to areas of different land use (agricultural, industrial or urban).

Naming habitats on the basis of physical features or the dominant vegetation is neither very systematic nor objective. Nor does it tell us anything about the plant community structure or species assemblages (groups of plants that have evolved together in a natural association). An alternative lies in groups of species that share similar ecological characteristics, for example xerophytic species (living in dry conditions). Attempts to identify roadside plant communities on the basis of formal and agreed-upon classifications are very few. In England, results of the Cheshire Roadside Verge Survey suggest two types of grassland based on the National Vegetation Classification.

The difficulty with applying plant community classifications to roadside verges is that many verges have been disturbed and new species have colonised alongside original species. In New Zealand, Wilson et al. (1992) after looking at distributions and climatic correlations of some exotic species, concluded that the characteristic species required for traditional classifications were too few. They further conclude that this paucity indicates a lack of constancy in the assemblages. Factors contributing to this lack of constancy could be geographical variation in roadside conditions as well as modified geology and soils of road shoulders.

Of the few surveys of roadside vegetation communities, those by Lausi and Nimis (1985) and Ullmann and Heindl (1989) are more detailed than most. Ullmann and Heindl analysed the general features and differentiation of roadside plant communities (mainly on the basis of coenosis or ecological preferences) in temperate Europe. They found that most of the roadside plant groups lacked definite associations (defined by classifications based on undisturbed plant communities). These plant communities were dominated by species from the families Poaceae, Compositae and Apiaceae.

Lausi and Nimis surveyed roadside vegetation along the major highways in boreal southern Yukon and adjacent Alaska in 1983. Using multivariate analysis, they established a positive correlation between the floristic composition of weed stands and adjacent natural vegetation. Three community types were identified for purposes of classification and seven predominant species groups (or coenoses, i.e., an assemblage with similar ecological preferences).

— Mesophytic species (living in intermediate conditions of moisture and temperature)
— Xerophytic species (living in dry conditions)
— Hygrophytic (living in wet or moist conditions)
— Coastal-montane oceanic
— Ubiquitous

Box 3.1: The Cheshire Roadside Verge Survey

Habitat types used in the survey. Note width of verge in relation to width of road.
— Deciduous woodland edge
— Wet heath
— Dry neutral heath
— Unshaded pond
— Dry Acid scrub
— Dry acid tall herb
— Wet acid tall herb
— Dry neutral tall herb
(Courtesy, Cheshire Wildlife Trust; photo, P. Hill)

— Xero-nitrophytic
— Anthropogenous (revegetation)

PLANTS

Examples of surveys of plants in roadside verges reported in the literature are shown in Table 3.1 along with general reports of roadside surveys.

Few plant species occur on roadside verges due to pre-existence before the road was built. Road construction and roadside embankments disturb the soils. Hence most of the species of plants occurring in roadside verges have been deliberately planted. Over many years replanting may amount to considerable equivalent areas. For example, in de Hamel's report of 1976, the author states that in the UK more than 8 million trees and shrubs were planted on major roads and motorways in the period 1967 to 1974. He then makes the curious observation that this is equivalent to 6000 acres (but fails to mention how this figure was derived).

Some plantings have colonised the verges while other plants have been accidentally introduced. Ruderal species are those which occur in disturbed sites among rubbish or debris (so often found on roadsides). In the USA, Federal highway networks provide new disturbed habitats (and migration routes) for the introduced purple loosestrife (*Lythrum salicaria*). A typical ruderal species in England is rosebay willowherb (*Epilobium angustifolium*). Some plant species, such as the pineapple mayweed (*Matricaria matricariodes*) in habit well-trodden tracks or well-used cart tracks. The sudden appearance of a 'sea' of red poppies on a new motorway embankment (Fig. 3.1) probably resulted from disturbance of a seed bank in soil transported to cover the new embankments.

In Britain, about 30 of the rarest plants are found in roadside verges and adjacent hedgerows (Royal Society for Nature Conservation, 1986). These include downy woundwort (*Stachys germanica*), Plymouth pear (*Pyrus cordata*), grape hyacinth (*Muscari atlanticum*) and field cow wheat (*Melampyrum arvense*).

An excellent example of a plant species inventory in the UK is the Cheshire Roadside Verge Survey (Cheshire Ecological Services, 1995). The aims of this survey were to review and revise a previous list of verges of importance for nature conservation and to prepare a working document for use by the Highways Department of the Cheshire County Council Engineering Department on the management of roadside verges (Box 3.1). The inventory includes 179 species of plants.

Approximately 800 of the 1400 species of plants found in the Netherlands can be found in roadside verges and even more along railway lines. About 160 species are found almost exclusively in roadside verges (van Bohemen, pers. comm.). Road verges are therefore an integral part of the diversity of habitats in the Netherlands.

Fig. 3.1. Sudden appearance of poppies on a new motorway embankment due probably to disturbance of a seed bank in soil used to cover the site (photo, Ian Spellerberg).

An interesting publication compares the flora of unimproved versus metalled roads (asphalted) in Belgium. Zwaenepoel (1997) found that 124 species were significantly associated more with unimproved roads (many of the species were of conservation interest). It would seem therefore that the indirect effects of metalled roads are more deleterious than such direct effects as pollution.

Trees planted in road verges are usually selected for their safety, visual and aesthetic properties (see Chap. 2). Secondary considerations might include the wildlife that would inhabit such trees. So, what wildlife do they attract? Most studies have centred on birds. According to Michael (1986), the relative use by songbirds of trees planted along interstate highways in West Virginia, USA amounted to only 10% for coniferous trees and a low 2% for deciduous. The most preferred tree species was cedar (*Juniperus* spp.). The species of tree is but one of many factors determining the abundance and diversity of bird species. Tree age, population density, associated plant species and extent of vegetation structure are also contributory factors

Weeds (a subjective term for those plants appearing in the wrong place at the wrong time) have been the subject of some roadside ecological studies and several authors have commented on the close links between climatic regions and weed distribution. In addition to the not unexpected comment that variation and diversity in roadside vegetation depend greatly on variation in soils, climate and altitude, a commonly mentioned topic is the relative abundance of indigenous versus exotic species. For example, introduced plant species accounted for 74% of the roadside flora on the slopes of the Hawaiian Mauna Loa volcano in 1982 (Wester and Juvik, 1983).

Larger and more prolific plants on roadsides have been found among the spectacular banksias in Australia. For example, Bryon Lamont and fellow workers found that specimens of *Banksia hookeriana* on roadsides were larger and more fecund (a previously unrecognised fact) than those growing elsewhere (Lamont et al., 1994). The crowns were more than twice larger and the flowerheads more profuse. Apparently, the absence of such complicating factors as weeds or fire pioneer plants and greater access to water enable such differences. Since commercial picking of this species on roadsides is prohibited, such populations may become critical for conserving genetic diversity within the species.

Other research in Australia has centred on the effects of roads on roadside vegetation. For example, it has long been known that in many parts of the continent mistletoes (Loranthaceae) are more abundant along roadsides than in adjacent vegetation. For some species the number of mistletoe plants per potential host tree may be tenfold that found in adjacent areas. Norton and Smith (1999) analysing why roadside mulgas provide better mistletoe hosts, concluded that roads exert a strong influence on the adjacent biota in arid central Australia. This influence occurs primarily through facilitation of water infiltration at roadside sites. In other parts of the world, water infiltration into roadside sites may be poor.

INVERTEBRATES

Examples of invertebrate surveys on roadside verges are shown in Table 3.2. Although contributing to habitat fragmentation, roads or at least roadside verges may provide linear habitats or refuges for some beetle species. Eversham and Telfer (1994) studied the beetle fauna (in particular, those beetles associated with open habitats of heathlands and sandy grasslands) of some roadside verges in the Netherlands and the east Anglian region of the UK. They found that roadside verges serve as refuges (rather than corridors). For those species dependent on early successive habitats, the chances of survival may be better on disturbed roadside verges than in unmanaged heathland nature reserves.

Ants are considered good indicators of the state of the terrestrial environment and any changes in it. Changes in species composition and structure of ant communities have been linked to environmental changes. In Australia, ant communities alongside roads were studied by Keals and Majer (1991). In South Africa, Samways et al. (1997) undertook surveys of species composition and abundance of ants in relation to roadside conditions on urban highways. Both studies highlight the importance of native vegetation, but findings differ with respect to widths of the roadside verges. Overall, the greater the width,

Table 3.2: Examples of the literature on invertebrate surveys in roadside verges in Europe

---

Braun, S. and Fluckiger, W. 1984a, b. *Surveys of aphids on hawthorn along motorways near Basle.*

Feltwell and Phillip. 1980. *The natural history of the M40 motorway.*

Free, J.B. et al., 1975. *Insects seen on flowers on motorway grass verges in the UK.*

Haeck, J. et al., 1980. *Colonisation of road verges by ground beetles in low-lying areas (polders) in the Netherlands.*

Munguira, M.L. and Thomas, J.A. 1992. *Surveys of road verges in southern England for butterflies and burnet moths.*

Olthoff, T. 1986. *Insect fauna on the most common trees in streets of Hamburg (in German).*

Port, G.R. and Thompson, J.R. 1980. *Surveys of insects on trees and shrubs used for landscaping major roads in the UK.*

Port, G.R. and Spencer, H.J. 1987. *Auchenorryncha (plant-sucking bugs) on roadsides in the UK.*

---

the greater the diversity of habitats. More heterogeneity is likely to encourage higher levels of species diversity.

The grass verges of some roads support populations of beneficial insects—pollinators and predators of agricultural and horticultural pests. In one survey of a motorway in the UK (M1), Free et al. (1975) found many beneficial insects and a few harmful to crops regularly visiting flowers. In regions where intensive agricultural practices have left few seminatural habitats, roadside edges, if properly managed, could play an important role for beneficial insects. Beneficial insects may predominate in some regions but pests in others. In Switzerland, Braun and Fluckiger (1984a, b) found hawthorn (*Crataegus* spp.) along motorway edges highly infested with the green apple aphis (*Aphis pomi*). Several possible reasons for these infestations were adduced, among which the following appeared paramount:

- Reduced efficiency of predators caused by the microclimate of the busy road.
- Biochemical changes in the host plant caused by motorway conditions and use of de-icing agents.
- Microclimatic effects, including slightly elevated temperatures.

Infestations of aphids and outbreaks of other insects are not unknown on motorways. Outbreaks of defoliating insects on trees and shrubs along major roads in the UK has prompted some interesting research. Many factors contribute to these outbreaks, including isolation from predators and pollutants that reduce microbial control agents. Port and Thompson (1980) commenting on previous observations that plants on motorway environments are frequently physiologically stressed, note that this quite likely increases nitrogen availability. Presumably, the nitrogen sources from NO_x pollution contribute to relatively high levels of nitrogen uptake in plants. This increase in available nitrogen in vegetation near major roads could very well promote increase in insect herbivore populations, in particular aphids.

One of the very few detailed studies of butterflies and moths on roadside verges was undertaken in southern England (Munguira and Thomas, 1992). They surveyed roadside verges and central reservations of 12 roads during the summer and found 27 species of butterflies (47% of the butterfly species in the UK and 56% of those in Dorset and Hampshire). The range of habitats on the verges affected the number of species and diversity (Shannon-Weaver diversity index). The density of adults and number of species correlated with verge width and diversity correlated with level of nectar abundance.

AMPHIBIANS AND REPTILES

Why didn't the toad cross the road? This question has been posed in the literature. True, roads can be major barriers for amphibians (see Chap. 4) but some aspects of roads are used by amphibians. Water-filled ditches and other bodies of water alongside roads can provide permanent or seasonal habitats for certain species of amphibians. But information on amphibian habitats and roads is scant. Contrarily, much has been published on the barrier effects of roads vis-à-vis amphibians.

Some species of reptiles are known to use road verges as habitats (Fig. 3.2a) and indeed in some countries the road verge is valued as a habitat for certain reptile populations (Wells et al., 1996). The way reptiles regulate their body temperature explains in part their use of road verges and the availability of fauna at the edges is an important incentive. Reptiles are ectothermic (as are invertebrates, fish and amphibians), that is, they depend on heat from the environment (not from their metabolism) to achieve body temperatures suitable for normal activity. Some small reptilian species commonly move from shade to sun and back to shade to regulate their body temperature. Along some road verges there is a marked transition between vegetation and bare surfaces. The physical conditions and microclimate of these edges thus provide in some instances a microenvironment wherein reptiles can bask on open ground, then retreat to shade when their body temperature has reached the requisite level. Vegetation also provides shelter from predators.

The road verge may also support a greater variety of food items for some reptiles. Certain species of spiders and insects may be more abundant along edges.

BIRDS

Birds are associated with roads for many reasons. Some feed on insects and other invertebrates stunned by vehicles. Carrion feeders such as kestrels are comonly sighted along motorways. Crows frequent motorway edges in North America (Sherburne, 1985) and in Britain. Typically, crows are carrion feeders in Britain but the curious observation has been made that they could be attracted to motorway verges because traffic vibrations cause earthworms to

Fig. 3.2a. Lizards basking on the roadside.

Fig. 3.2b. Bird nests in a roadside transformer (photos, Ian Spellerberg).

exit their burrows and emerge on the ground surface (Tabor, 1974). Aside from serving as food sources, roadside verges may provide habitats for many species of birds. In the USA, some hawk species use roadside verges during winter when southerly migrations take place.

Birds use many of the structures associated with roads. For example, pigeons commonly use bridges for roosting and nesting. In northern Maine, USA, barn swallows, cliff swallows and rock doves have been found nesting beneath bridges (Sherburne, 1985). Birds commonly settle on power lines and their nests are sometimes found in transformers (Fig. 3.2b) on power lines.

What bird species frequent motorway verges and which are the most common and why? These may appear to be rather naïve questions and many bird specialists (given information on the biogeographical region) would readily be able to give intelligent answers. But despite the variety of reports on birds found along or near roads (Table 3.3), the basic ecological questions of which and why have not been well surveyed. In a 6-year study on the motorways around Brno (Czech Republic), Havlin (1987) recorded 82 species (of which, 73 were identified). The most frequently occurring species were *Aluda arvensis* (skylark) and *Phasianus olchicus* (pheasant). Other species were *Perdix perdix* (partridge), *Saxicola torquata* (stonechat), *Falco tinnunculus* (kestrel) and *Buteo buteo* (buzzard). Other commonly observed species included pigeons, *Columba livia domestica* (mostly flying past) and *Passer domesticus* (sparrows) found mostly near bridges and housing estates. Different levels of abundance were found at different localities and between motorways and motorway junctions. Havlin's study is the first step towards explaining what determines species composition, abundance and diversity. Additional helpful factors would be the variety of habitats (for feeding, roosting and nesting), use of the surrounding land, local climate and vegetation. De-icing salts also attract some birds (*Loxia curvirostra* (crossbill), *Columba livia domestica*) and some species feed off carrion and invertebrates on the road surface.

Width of roadside verges, management of verges and the surrounding land use may all be important factors determining which birds frequent road sites and their abundance. In Illinois, USA, for example, Richard Warner

Table 3.3: Examples of the literature on surveys of birds in roadside verges

Australia
Newbey and Newbey. 1987. *Use of road reserves by birds in Western Australia.*

Europe
Havlin, 1987. *Surveys of birds along motorways in Czechoslovakia.*

Scandanavia
Laursen. 1981. *Use made of roadside verges by birds.*

USA
Michael. 1986. *Nesting frequencies of songbirds in West Virginia.*
Warner. 1992. *Ecology of nesting by grassland passerines along roads in Illinois.*

studied the nest ecology of grassland passerines in rural areas. He found that the number of nests and species of passerines increased in most cases with roadside width (Warner, 1992). High nest density in some areas reflected sparse prime nest cover in surrounding farmland and frequency of nesting was further affected by grass species and mowing frequencies.

Arnold and Weeldenburg (1990) undertook a census of birds on road verges in the wheatbelt of Western Australia (Fig. 3.3). Twelve surveys were undertaken per annum. The abundance of 20 of the 21 most common species correlated with road-verge characteristics and the authors concluded that width of verge and area of native vegetation significantly affected the density of several species.

MAMMALS

Studies of small mammals in roadside verges and effects of verge management on them are numerous. As long ago as 1941, Laurence Hughey published an account of mammals (Table 3.4), in particular how pocket gophers (*Thomomys*) extend their range via roads. Since Hughey's work, studies on dispersal of small mammals and the possible role of road verges as corridors are surprisingly few.

Adams and Geis (1983) undertook a major study of the effects of roads on species richness, spatial distribution and population density of small mammals in three geographic areas of the USA (southern piedmont of Virginia and North and South Carolina, midwest tillplain of Illinois and the valley region of Oregon). They found 40 species of small mammals; the most prevalent species were the white-footed mouse (*Peromyscus leucopus*) and the deer mouse (*Peromyscus maniculatus*).

Compared to mammal studies on roadside verges, similar studies on highway median strips are scant. Adams (1984) studied the effects of habitat type on small mammal use of a highway median strip in the Piedmont plateau region in North Carolina, USA and compared the results with adjacent roadside verge use. The median strip included a wooded area about 93 m wide bordered by an unmown strip (right-of-way) and then a mown strip (right-of-way). Mammal density proved similar in both median strips and adjacent verges and some species were common to both areas. The difference in species composition could only be explained with further study.

MANAGING THE VERGES

Road verges (and right-of-ways) are managed for a variety of reasons: road safety, drainage, soil stability, fire risk, stock feed, weed control, control of stock movements, people access, aesthetics, landscapes and conservation of biota. They are generally kept free of shrubs and often subjected to frequent mowing. Management for aesthetic purposes is often a main secondary

58 Ecological Effects of Roads

Fig. 3.3.

Fig. 3.3. Wheatbelt landscape of Western Australia and photograph of a roadside edge showing few trees (photos, G.W. Arnold).

Table 3.4: Examples of the literature on surveys of mammals in roadside verges

Australia
Bennett, A.F. 1988. Roadside vegetation and mammals at Naringal, southwestern Victoria.

USA
Adams. 1984. *Surveys of small mammals in road verges in Illinois.*
Adams and Geis. 1983. *Survey of small mammals along roads in Illinois and Virginia.*
Getz et al. 1978. *Survey of meadow voles in road verges in Illinois.*
Hughey. 1941. *Surveys of pocket gophers along roads in south-western USA.*
Oxley et al. 1974. *Survey of small mammals along roads in Ontario and Quebec.*

objective. In the past, management for conservation was a low priority. The cost of management can be reduced by establishment of wildflower swards as these require less maintenance and less mowing. Wildflower swards are attractive and the grass verges look 'tidy'.

Howell's book *Roadside Bio-Engineering* published in 1999 is an excellent reference manual on the role of vegetation in management of roadside verges from an ecological engineering perspective. Although written mainly for the environment in Nepal, this manual does include some important basic principles about management of roadside vegetation in general.

The ecological importance of road verges is not only recognised in Schaffer's Ph.D. thesis *Ecology of roadside plant communities* accepted for publication in 2000, but also provides a very detailed ecological framework as a basis for the establishment and management of road verge plant communities. Pertaining mainly to roads in the Netherlands (where it is estimated that roadside habitats make up almost 2% of the total land area), Schaffer provides perhaps the very first detailed and comprehensive ecological basis for roadside plant communities.

In 1986, the Royal Society for Natural Conservation (RSNC) (UK) published a pamphlet *Wild flowers on the verges* as a guide for users and managers of roadside verges (Box 3.2). The pamphlet describes the origins of roadside verges (many roads were constructed between 1750 and 1850) and comments on ownership and the law (most verges are part of the highways but some are common land). It also provides a management code (Box 3.2), one of the few examples of readily available advice for roadside verge management. Lastly, the pamphlet suggests how various organisations and individuals can assist in management and conservation of wildflowers on roadside verges. Organisations to contact in the UK include Local Nature Conservation Trusts, Nature Conservancy Council (English Nature, Scottish Natural Heritage, Countryside Council for Wales), and regional and local councils.

The 1994 English Nature publication *Roads and Nature Conservation* (Ramsay, 1994) provides some very good case studies of overall mitigation and enhancement of habitats in the wake of road projects (Box 3.3 below; Chap. 7, Box 7.3) as well as some brief suggestions for road verge manage-ment.

Management methods for vegetation on roadside verges can range from mowing through grazing to the use of herbicides. Some regional authorities in southern England such as Hampshire County Council have introduced roadside verge mowing regimes to cater to the requirements of certain butterfly species. Management with herbicides has certain risks. The selective actions of herbicides are likely to cause loss of vegetation structure, i.e., the structure or complexity of the vegetation is likely to decline. The risk of toxic effects on animals must also be kept in mind.

One particular challenge facing managers of roadside verges is how to combine management of introduced plant species with management of naturally occurring species. For example, are the management practices

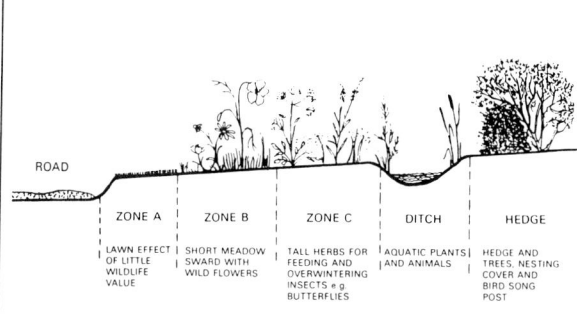

```
ROAD    ZONE A    | ZONE B    | ZONE C    | DITCH    | HEDGE
        LAWN EFFECT| SHORT MEADOW| TALL HERBS FOR| AQUATIC PLANTS| HEDGE AND
        OF LITTLE | SWARD WITH | FEEDING AND   | AND ANIMALS   | TREES, NESTING
        WILDLIFE  | WILD FLOWERS| OVERWINTERING|               | COVER AND
        VALUE     |            | INSECTS e.g. |               | BIRD SONG
                  |            | BUTTERFLIES  |               | POST
```

THE MANAGEMENT CODE

If you manage roadside and any other wayside verges, follow the Management Code below.

Verges are historically a managed habitat and it is this that has given rise to their value for wild flowers. To stop cutting will result in the development of scrub and trees which will reduce the great variety of grassland flowers (and the associated butterflies and other insects) that make road and other wayside verges so attractive.

Management of roadside verges

Cutting A complete absence of cutting can be as detrimental to certain low-growing species of wild flowers as is intensive cutting. In order to produce verges that are of real wildlife and visual value it is important to create a range of different habitats across the width of the verge as illustrated below.

Zone A
1m adjacent to the road to be cut as necessary for road safety with a maximum of six cuts, the first at the end of May, keeping the grass about 20cms — avoid 'skinning' the turf as this encourages annual weeds.

Zone B
The next 1-2m to be cut — twice per year. The first in June and the second after the end of August. If cowslips predominate, cut high in June to spare their seed.
or — once per year in late August.

Zone C
Next to the hedge or ditch, to be cut once every two years in late August or not at all. Preferably alternate sides or stretches every other year.

Chemicals Use of chemical sprays will severely reduce the variety of wild flowers on the verge. Herbicides should be used only in exceptional cases, to deal with 'injurious weeds' for example.

Weeds There are five 'injurious weeds' defined in the Weeds Act 1959. These are:
Broad-leaved Dock
Curled Dock
Creeping Thistle
Spear Thistle
Ragwort
Under the Weeds Act, the Ministry of Agriculture, Fisheries and Food has powers to require an occupier of land to prevent the spread of these weeds.

Thistles on verges can often be the biggest problem to farmers, particularly where the verges have been disturbed by spreading spoil, ploughing or vehicular use. They can be controlled by cutting annually just before the flower buds open and by protecting the verge from damage.

Management of special verges

This applies to road verges that have been designated as Sites of Special Scientific Interest (SSSI), by the Nature Conservancy Council, or noted as nature reserves or special interest areas by the local Nature Conservation Trust.

Special management plans should be drawn up for such areas and regular liaison should take place between the designating body and the managers. In the case of Sites of Special Scientific Interest, there is a legal obligation to consult the Nature Conservancy Council before undertaking any operations, with a penalty of up to £1,000 for failure to consult.

Management of roadside features

Trees & Tree planting Choice of tree type is important. Native species that grow naturally in the area are likely to take best, and will benefit wildlife most. Permission is required from the Highway Authority to plant trees on roadside verges.

Consult the County Council for advice and grant aid. Check with the local Nature Conservation Trust or the Nature Conservancy Council before planting on special verges.

Hedges are best cut in the autumn or winter when the disturbance to birds and other wildlife is less. However, cutting at this time of year can remove essential food supply for birds in the form of berries. Ideally then, hedges should be managed in rotation so that a good local food supply for wildlife is retained.

Ditches are particularly valuable wild flower and landscape features. Wet ditches should be cleaned out on a rotation of 3 to 8 years, the larger ones needing attention less frequently. Ideally, small consecutive stretches would be done each year, giving a variety of stages of colonisation by aquatic plants and animals.

Dry ditches also have their value if tall fringing vegetation is allowed to grow up. Cutting alternate sides every other year will keep the continuity for wildlife.

Management is best done in the autumn and winter when disturbance is generally less.

Care should be taken that disposal of spoil does not smother roadside verges. Ideally it should go back on the fields.

Management of newly created verges

Seeding with a low maintenance grass species such as bent or fescue will produce a sward which will allow wild herbs to seed in and co-exist. Where such herbs and hence their seed are scarce, the inclusion of commercial wild flower seed should be considered and advice sought on suitable mixtures to recolonise.

Lost hedges should be replaced and new ones should be established wherever possible and hedgerow trees encouraged using species native to the area.

Box 3.2: Excerpt from the RSNC pamphlet *Wild flowers on the verges*

appropriate for the former compatible with those for the latter? How do introduced plants interact with the naturally occurring ones? The current literature carries few answers to these questions and it appears that the ecology of plants in the context of roadside verge management is an area requiring further research.

POTENTIAL FOR NATURE CONSERVATION

In many parts of the world and especially in areas of high population density, roadside verges are one of the few remaining areas not only for supporting wildlife communities, but also with potential for conservation.

Box 3.3:

Management of grass swards, from *Roads and Nature Conservation; guidance on impacts, mitigation and enhancement. English Nature.*

Grassland creation

Grassland can be established in a number of ways:–

- Natural colonisation
- Hay collected from a nearby field in July-August, when the required seed is still on the plants, can be strewn over the surface. If applied thickly, the material needs to be removed after about 3–4 weeks after seed has been shed or it will smother the seedlings. Laid more sparingly, there will be less seed, grassland establishment will be slower, but the hay does not need removing. In the latter case, bare ground should be visible in a small-scale patchwork through the hay.
- Seed can be collected from a nearby hay field by using a non-destructive brush-collector feeding into a collecting box. By collecting in July, August and September, and mixing the resulting harvest, late and early flowering species can be acquired.
- A hay field can also be combined using traditional agricultural equipment, but set to work slowly with a low fan speed. Only the seed ripe at the time of harvest will be available, but missed, early flowering species can be added.
- By making local arrangements with, for example, English Nature, a Country Wildlife Trust or local landowner to collect from a Nature Reserve or SSSI, significant savings in seed costs can be made. For instance, hay seed used in one roadside scheme cost £15/kg.
- Using a wildflower seed mix will be more expensive, but is likely to fall within the range £370 to £960/ha at 30kg/ha.
- Seed mixes are usually 80-85% grasses by weight, with the remainder broad-leaved herbs. The proportion will depend on the weight of different species-differences between the largest and smallest seed weights are considerable, for example, the bent grass, *Agrostis castellana* has 10,000 seeds/g, whilst tufted vetch *Vicia cracca* manages only 40. The high proportion of grasses keeps the cost down, but semi-natural grasslands can consist of 60% or more grasses.
- Seed mixes are usually sown at about 30kg/ha, but this needs to be increased where seed is hydroseeded, and where hay seed mixes contain more chaff. A rate of 40kgs/ha has been found to be adequate, even on step slopes, using the latter. Seed is best sown in autumn, except at high altitudes or on exposed sites.
- Establishment techniques need to follow the supplier's guidelines.

Contd.

> **Box 3.3 Contd.**
>
> - There is much conflicting advice on managing grasslands after establishment. The actual needs are much more simple than is often portrayed. In the first growing season up to 4 cuts, with the arisings removed if possible, need to be allowed but may not be required. Cutting is necessary only if the growth of the first species to establish are threatening to overtop those still germinating. On some subsoil sites no cutting has been necessary for up to 2 years after sowing. If topsoil has been used, the full number of cuts will probably be required. However, poor growing conditions, as in cold springs or droughts, avoids the need for some cuts.
> - After establishment, grasslands are best managed to encourage a wide range of flowers and fruits, to last for as long as possible. This will benefit invertebrates in particular, and provide long-term cover for small mammals.
> - The edges of the grasslands, where they intermingle with shrubs and adjoin a hedge or woodland, are best left uncut, or cut only periodically, thus providing hibernation tussocks for invertebrates.
> - The rest of the grassland will need cutting and the arisings removed to prevent the litter from blanketing out the more delicate species. On subsoil, management can be restricted to once a year, or possibly once every 2 or 3 years or less regularly, in September-October after the invertebrates have mostly completed their life cycle. On moderately vigorous swards on thin topsoil or more productive subsoils, a single cut and clear in late autumn should suffice. Where grasses are threatening to dominate at the expense of less vigorous plants, the sward needs to be cut close and the arisings cleared in late March or April. This removes any winter's growth and reduces the vigour of the grasses. The timing is critical, since it should avoid the main bird breeding season and the emergence of most of the invertebrates.
> - The promotion of spring and summer meadows, as described by others, is unnecessarily complicated and very damaging to the zoological interest of the grassland. It is an approach developed principally for different user seasons in urban parks and is not appropriate to roadside grasslands or the creation of valuable wildlife habitats.

Whereas in the past road construction led to reduction in wildlife habitat and disturbance of wildlife populations, some new roads (particularly the road verges) do provide opportunities for conservation of some forms of wildlife. In Australia for example, the Victorian Conservation Strategy (1988) identified roadside management plans as a measure for managing and protecting native vegetation on roadside verges (Straker, 1998). The Roadside Conservation Committee (of Victoria) is at the forefront of achievement in promoting conservation values of roadsides and in protecting roadside habitats of native flora and fauna (Gilbert, 1998).

In the Netherlands, the ecological role of roadside verges is also well recognised by the government as well as road authorities. This recognition applies both to plant communities and certain other taxa. In the UK, Thomas et al. (in press) have concluded that carefully designed areas along new roads do provide considerable opportunities for enhancing rather than depleting butterfly populations—provided the road route lies on land which generally lacks wildlife. The authors note, however, that whereas the habitat creation beside the M40 and around M3 on Twyford Down has successfully increased levels of wildlife, this does not justify laying new roads through natural or seminatural areas rich in wildlife.

When the 12-mile long King's Langley and Bethamsted bypass in England was opened in 1993, press coverage of the longest wholly deliberate linear nature reserve abounded. Almost 80,000 indigenous trees, 15,000 shrubs and many other plants together with hundreds of kilos of wildflower seed were used in establishing the roadside verges for nature. Fences and tunnels were also incorporated in the road design. While some welcomed this restoration, there were those who held that it merely fiddled with the margins of the road and did little to compensate for the devastation caused by construction of the bypass. Such linear habitats are welcome, said others, but their contribution to mitigation of road effects is insignificant.

New roads may provide opportunities to establish conservation values of roadside verges by way of habitat restoration. New roads may also provide opportunities for establishment of wildlife corridors between fragmented habitats with isolated populations. However, research is scant on what is needed to make wildlife corridors functional for movement between populations. Furthermore, the corridor function can also be negative. In Yellowstone National Park for example, roads enable bison to leave the protection of the Park and hence risk being hit by vehicles (Marnie Criley, pers. comm.). A recent book edited by Harper-Lore and Wilson (*Roadside use of native plants*) provides an excellent list of native plants suitable for planting on road verges throughout states in the USA.

Some plans for habitat construction have been drawn up but never implemented. For example, during construction of the M3 motorway through Twyford Down, England (Winchester bypass; Chap. 1, Fig. 1.2) concerns were expressed by archaeologists and local community groups worried about visual impact. A concomitant plan to re-establish chalk grassland (Morris et al., 1994) was not immediately implemented despite detailed research on the ecological principles and processes involved. Some years later, however, a chalk grassland nature reserve was established about 1.5 km away from the motorway. Several chalk grassland butterflies, including the rare small blue butterfly (*Cupido minimus*), are now resident there (Barry Fox, pers. comm.).

Roadside verges may provide opportunities for conservation of some wildlife. Road embankments in chalk grassland areas of southern England provide habitats for butterflies. But what appears to be a simple conclusion may be difficult to put into practice. Difficulties arise in part from different and sometimes competitive interests in management objectives for roadside verges. They also arise from parties with vested interests. Management of roadside verges for conservation requires more than ecological knowledge; it also requires a process for achieving communication and co-operation between interested parties. These parties include those directly responsible for road-verge management (highway or road maintenance authorities), government conservation agencies and non-governmental conservation agencies such as naturalist trusts. All these potential hindrances have implications for better communication between the interested groups, integration of values and ideas and, last but not least, for databases on what is present in a roadside verge, its condition, how it is managed and by whom.

CONSERVATION PRIORITIES

Some roadside verges could usefully be managed for conservation. Some support rare species and/or rare or endangered plant communities. The question is: How to ensure recognition of such roadside verges as important contributors to conservation? Faced with the limited resources for management of roadside verges, it may be necessary to establish a structured process for identifying conservation priorities. In other words, define the criteria to be used in assessing the conservation value of a roadside verge.

Many ecological evaluation methods are available for determining conservation priorities (Spellerberg, 1992) but few explicitly pertain to roadside verges. The Cheshire Roadside Verge Survey identified the following criteria for determining whether a verge could be of importance in nature conservation:

— Typicality
— Rarity
— Assemblage
— Aesthetic appeal
— Importance as food/nectar source for insects

The criteria typicality and rarity are included in the ten criteria considered priorities by Ratcliffe (RSNC) in the designation of a nature reserve.

Identifying conservation priorities is important in the planning of routes for new roads. The method developed by Yapp and published in 1973 was probably the earliest attempt to introduce a structured process for identifying routes least likely to affect conservation priorities when profiling routes for new roads. Numerical scores are derived from a classification of land types and the length of each type along the road (Table 3.5). This method is fully described and assessed by Spellerberg (1992).

Table 3.5: Two examples (Route A and Route B) of Yapp's method of determining the extent of damage along alternative routes for a new road (modified from Spellerberg, 1992)

	(1) km	(2) Type	(3) Features	(4) Class	Units of damage (Columns 1 x 4)
Route A	7	Agricultural land	1+2+1+3+1+1=9	2	14
	5	Planted woodland on one side, lake with secondary regeneration and foreshore on the other		4	20
	1	Existing new road		0	0
	1	Urban		0	0
	2.5	Good semi-natural deciduous woodland, in part a scheduled site of special scientific interest		5	12.5
	3	Existing new road		0	0
	3	Part planted woodland, part bog		4	12
	6	Agricultural land	2+3+2+0+0+2=9	2	12
					70.5
Route B	15	Agricultural land	1+1+0+1+1+0=4	1	15
	17	Agricultural land	3+2+3+1+1+0=10	2	34
					49

4

Habitat Fragmentation, Barriers and Corridors

INTRODUCTION

Over thousands of years, natural processes (fire, floods, earthquakes, changes in river flows) have caused habitat fragmentation. However, many human activities also cause habitat fragmentation. These include agriculture, afforestation, railway lines, canals, urbanisation and roads and road networks. By far the greatest effect of roads on nature occurs when new roads are built in natural areas or in previously or relatively undisturbed areas. Compared to a new road in a rural landscape, the impact of a new road in a natural or semi-natural area has long-term, large-scale and complex consequences for wildlife.

This Chapter deals with the phenomenon of habitat fragmentation—its description and analysis. The general effects of habitat fragmentation caused by roads and road networks are discussed (railway lines also contribute to habitat fragmentation—see for example van der Grift and Kuijsters, 1998). The account of the general effects is followed by a brief review of habitat fragmentation by roads and the effects on specific groups of plants or animals. In particular, roads can create barriers that make it difficult for some forms of wildlife to cross over from one side to the other. How some roads help to disperse wildlife and the role of road verges as linear habitats or greenways is addressed.

Lastly, what can be done to minimise the effects of habitat fragmentation and barrier effects is explored. In the last section of this Chapter, some mitigation methods are discussed (see also Chaps. 7 and 8).

HABITAT FRAGMENTATION:
GENERAL EFFECTS AND ANALYSIS

The places where wild plants, animals and other organisms are found (in the sea, lakes and rivers or on land) are many and varied. The term habitat is used as a general term for such places. Where a habitat ends and begins (at a particular point in time) is not always easy to identify. In some cases the extent of a habitat can be identified by a predominant physical feature or predominant vegetation characteristic. For example, a lake habitat (at a particular point in time) ends at the edge of the water. A woodland habitat is defined by the presence or absence of trees.

The area and shapes of many habitats can be measured if, indeed, it is possible to identify where the habitat ends. All habitats can, in a simple way, be divided into two general areas; the boundary (edge or ecotone) and the area enclosed by the edge (See Chap. 2.2). Both the edge or ecotone and the inner area or core area of a habitat will each have typical kinds of species.

Habitat fragmentation (sometimes referred to as landscape fragmentation) is a process during which wildlife habitats are divided into smaller and smaller components (Fig. 4.1). Not all habitat fragmentation is caused by roads but nevertheless the general research on the effects of habitat fragmentation is relevant to any appraisal of the ecological effects of roads.

The process of habitat fragmentation also increases the length of the edges around the habitat. Imagine a habitat that perfect circle. Such a shape has the minimum length of edge and maximum core area. Cut the circle into two halves and the core area of the two halves remains the same as the whole circle but the combined length of the edges of the semicircles (the area of the ecotone increases). Habitat fragmentation may present advantages for edge species or ecotone species but not for those species inhabiting the core area of a habitat. This is because there is less habitat available and because for some species the areas between the fragments may act as barriers preventing their movement from one fragment to another.

Many examples of habitat fragmentation can be seen in Europe and North America. For example, in Fig. 4.1 the different stages show the gradual loss of forests in a part of England and the gradual fragmentation of those forests until all that is left are a few fragments. Forests have been replaced by agricultural and urban developments.

Habitat fragmentation has become a major focus of research in conservation biology because many believe it has by far the greatest and most detrimental impact on nature. Wilcox and Murphy (1985), for example, claimed it to be 'the most serious threat to biological diversity, and the primary cause of the present extinction crisis'.

Several reviews of the literature dealing with habitat fragmentation have been published and some are listed in Table 4.1. The literature includes material on:

— roadside verges as habitats
— management of roadside verges for wildlife
— dispersal of wildlife along verges; possible corridor functions
— dispersal of alien and pest species.

A detailed overview of the effects of habitat fragmentation is shown in Table 4.2 (from Schonewald-Cox and Buechner, 1992). Habitat fragmentation includes progressive loss of habitat, fragmentation of remaining habitats caused by agriculture, forestry, urbanisation, roads and other linear man-made features and human disturbance.

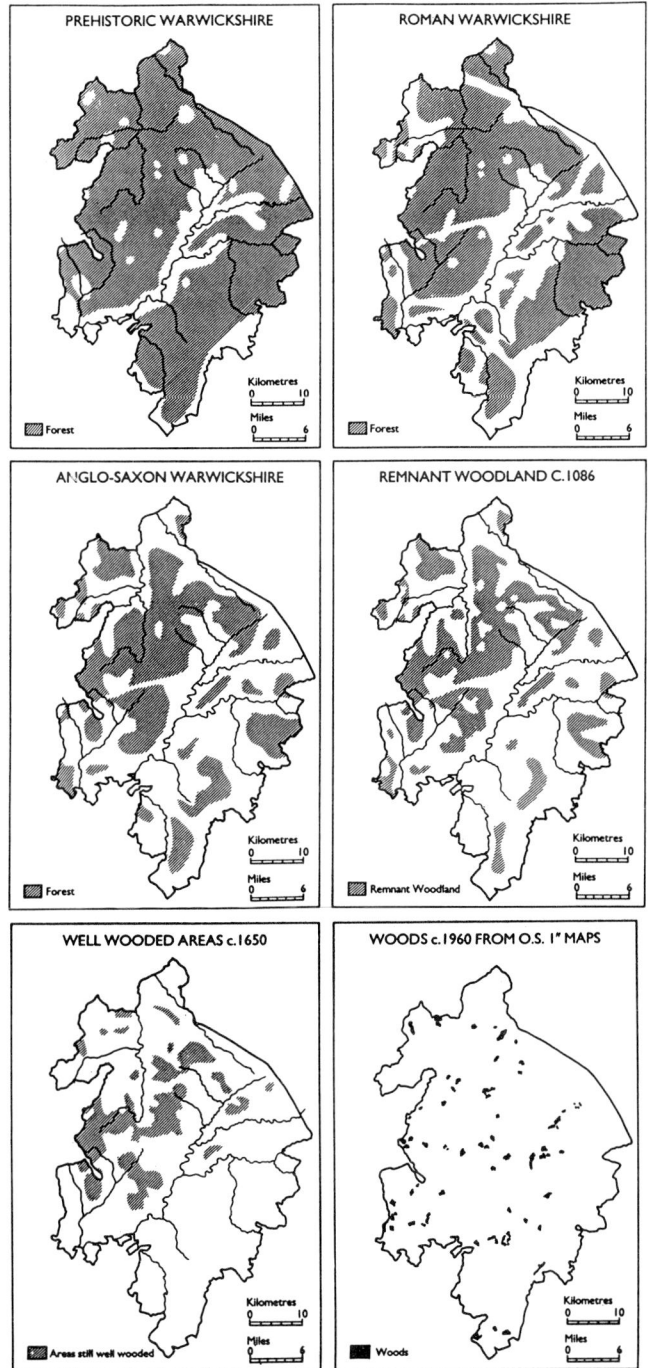

Fig. 4.1. Example of habitat fragmentation; over time, the habitats are reduced in area, fragmented and isolated. (Modified from Thorpe, 1978).

Table 4.1: Examples of (a) general literature dealing with habitat fragmentation and (b) literature on the effects of roads of habitat fragmentation.

(a)

Robinson, G.R. and Quinn, J.F. 1992. Habitat, fragmentation, species diversity, extinction, and design of nature reserves. pp. 223-248 In Jain, S.K., Botsford, L.W. (Eds.). Applied Population Biology. *Monographiae Biologicae*, Vol. 67. Kluwer Academic Publ. Dordrecht.

Spellerberg, I.F. 1991. Biogeographical basis of conservation pp. 293-322. In Spellerberg, I.F. et al. (Eds.). *The Scientific Management of Temperate Communities for Conservation*, Blackwell Science, Oxford.

Usher, M.B. 1987. Effects of fragmentation on communities and populations; a review with applications to wildlife conservation, pp. 103-121. In Saunders, D.A. et al. (Eds.). *Nature Conservation: the Role of Remnants of Native Vegetation*. Surrey Beatty & Sons, Australia.

Wilcove, D.S. et al. 1986. Habitat fragmentation in the temperate zone, pp. 237-256. In Soule, M.E. (Ed.). *Conservation Biology: the Science of Scarcity and Diversity*. Sinauer Associates, MA, USA.

Wilcox, B.A. and Murphy, D.D. 1985. Conservation strategy: the effects of fragmentation on extinction. *American Naturalist*, 125: 879-87.

Young, A.G. and Clarke, G.M. (eds.). 2000. *Genetics, Demography and Viability of Fragmented Populations*. Cambridge University Press, Cambridge.

(b)

Andrews, A. 1990. Fragmentation of habitat by roads and utility corridors: a review. *Australian Zoologist*, 26: 130-141.

Canters, K., Piepers, A. and Hendriks-Heersma, D. 1997. Habitat fragmentation and infrastructure. The role of ecological engineering. *Proc. International Conference, Maastricht*, the Hague. Ministry of Public Works and Water Management, Directorate-General for Public Works and Water Management, Road and Hydraulic Engineering Division. ISBN 90-369-3737-2.

Curzydlo, J. 1999. Ecological passages for wildlife and roadside afforestation as necessary parts of modern road constructions (Motorways and railways). *Proc. International Seminar*, Krakow, 7-10 September, 1999. ISBN 83-912184-1-4.

Schonewald-Cox, C. and Buechner, M. 1992. Park protection and public roads, pp. 373-395. In Fieldler, P.L. and Jain, S.K. (Eds.) *Conservation Biology*, Chapman & Hall, London.

Habitat fragmentation thus results in loss of habitat, increase in number of habitat fragments and for some species isolation of populations within species (which may have implications for loss of biological diversity at different levels of biological organisation). For some taxa, the area of habitat becomes too small to support the resources needed for survival. Populations of some species may become reduced in size and that has implications for biological diversity at the population level of organisation. Diaz et al. (2000) hypothesise that the combined effects of fragmentation and predation in small remnants of forests have led to the extinction of the forest lizard *Psammodramus*

Table 4.2: Summary of the main effects caused by habitat fragmentation, particularly those brought about by roads. (After Schonewald-Cox and Buechner, 1992) with permission of Kluwer Academic Publishers.

I. MODIFICATION OF HABITAT

A. Changes in shape and size of landscape elements
   1. Decreased size of continuous habitat in remnant patches
   2. Altered shape of continuous areas of patch interior habitat
   3. Altered geometry of edges
   4. Increased perimeter area ratios of remnant patches

B. Changes in connectivity and isolation of landscape elements
   1. Increased degree of isolation of remnant patches for species, material or effects restricted to patch interior habitats
   2. Increased connectivity of remnant patches for species, materials or effects following edge or modified habitat
   3. Increased access for logging, mining, hunting and other resource-extraction activities
   4. Increased access for poachers and other illegal activities

C. Changes in habitat types
   1. Increased amount of edge and modified habitats
   2. Decreased amount of patch interior habitats
   3. Changes in composition and geometry of edge habitats
   4. Loss of sensitive species from small remnant patches
   5. Altered balance of exotic and native species
   6. Altered balance of weedy or edge and patch interior species
   7. Increased spatial and temporal variation in habitat quality for patch interior species
   8. Increased habitat homogeneity within small remnant patches
   9. Changes in capacity of the reserve for populations of sensitive species

II. MODIFICATION OF QUALITY OF PROTECTION PROVIDED

A. Changes in balance of patch interior versus edge species and native versus exotic species

B. Increased exposure of internal areas and further subdivision of landscape
   1. Direct removal of habitat
   2. Increased amount of edge in landscape
   3. Increased exposure to edge effect
   4. Increasing fluctuation of microclimate and related processes
   5. Influx of foreign materials (pollen, insects, toxins, refuse)
   6. Disturbance of habitat

C. Declines of populations of species that
   1. occur naturally at low densities
   2. have large area requirements
   3. do not do well in edge habitats

Contd.

Table 4.2: (Contd.)

  4. are sensitive to human contacts
  5. are likely or unable to cross roads
  6. are frequently killed on roads (e.g., seek out roads for heat or food)
  7. are otherwise sensitive to extinction resulting from habitat fragmentation or disturbance

III. MAJOR OBSERVED CHANGES

A. Peninsular effects and some island effects
B. Altered population dynamics of many species
C. Possible increased probability of further fragmentation
D. Increase in absolute amount of edge in the landscape
E. Decrease in amount of edge that can support sensitive species
F. Subdivision of protected habitats and forced metapopulation structure of patch interior species
G. Altered patch dynamics; for example, loss of species for which patch colonisation rates are lower than local patch extinction rates
H. Increased instability of ecological processes and increased frequency of fluctuation in habitat quality
I. Predisposition of local extinction of some species

---

algirus in fragments smaller than about 90 ha. Recolonisation seems to be prevented by the very limited dispersal abilities of these lizards.

Forest fragmentation has been shown to affect the presence/absence of some bird species. More importantly, it appears that forest fragmentation may alter fundamental ecological processes that determine the nature of ecological communities. For example, Harris and Silva-Lopez (1992) in reviewing forest fragmentation in the north-central Florida region (dominated by the Ocla National Forest), identified five groups of ecological functions that resulted from fragmentation of habitats there, namely: increased levels of competition among birds for nesting cavities, increased levels of open-nest parasitism, increased levels of ground nest predation and increased levels of parasites and disease.

Reduction in area of biotic communities also has implications for ecological processes and resilience of communities in recovering from perturbations. This is because as the area becomes smaller, there is less flexibility. In some biotic communities, the occurrence of natural disturbance regimes such as gaps or cleared areas (ranging in scale from the small mounds of soil made by moles digging burrows to land-slips) and perturbations such as flooding and fire are important functions which vary in scale. Some biotic communities become so small and isolated that they are at risk of total destruction by fires or other major disturbances. The habitat has become so reduced in area that it restricts natural community mosaics.

In a paper reviewing the protection of natural areas in fragmented landscapes, Noss (1987), as have other authors, draw attention to the need for developing an integrated network of protected areas as a means of overcoming the effects of habitat fragmentation. In other words, rather than designating a few unconnected habitat fragments as protected areas, it is much better to think in terms of a series of protected areas inter connected via linear landscape features (some of which may be wildlife corridors). One proposed network of protected areas and corridors is that shown in Fig. 4.2 for the southern region of Ohio (USA). In 1987, Noss noted that idealistic land conservation projects are not as popular in Ohio as in other states, illustrating the importance of the perceptions of human communities and current politics.

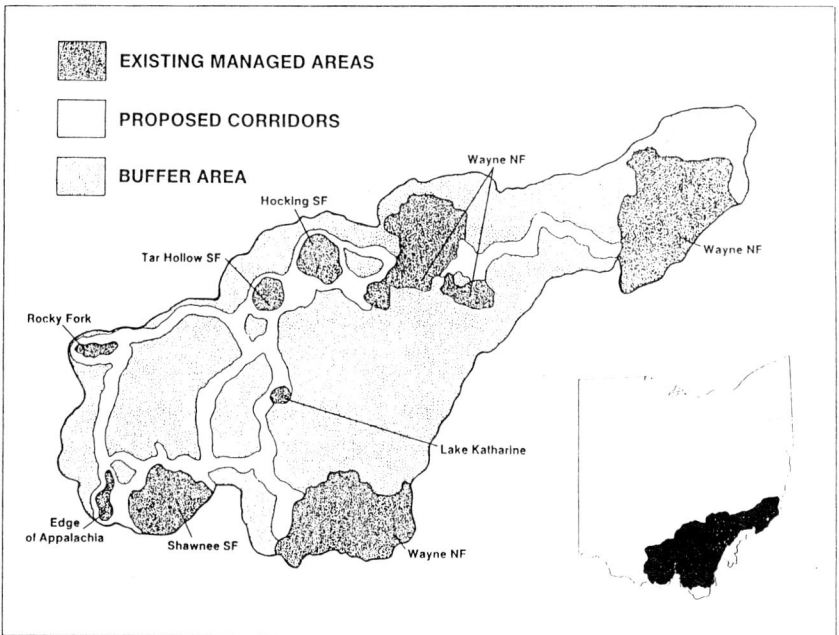

Fig. 4.2. Proposed protected area network of conservation areas for Southern Ohio, USA (from Noss, 1987).

That the overall effect of fragmentation is loss of wildlife cannot be disputed. Wide-ranging predators in particular are negatively affected by habitat fragmentation. However, such fragmentation can favour some species, in particular large birds and mammals which are wide ranging. Food resources for large birds of prey are likely to be more diverse and more plentiful if the feeding area is not too homogeneous (such as one continuous and uninterrupted forest) and comprises diverse types of habitats and different habitat areas. Put another way, a patchy environment may be favoured more by certain species. In plantation forests for example, the roads are generally the main cause of structural landscape diversity. It is along those roads that diversity of habitats is seen, which supports far more wildlife than the interior of plantation forests.

Before attempting to understand the effects of habitat fragmentation on biota, it is necessary to quantity the patterns and processes of fragmentation. That is, there needs to be some kind of analysis. There are many ways of analysing habitat fragmentation. The first question might be: Where does the habitat fragment start and end? In addition to plotting the rate of reduction of total habitat area and rate of increase in number of fragments, studies on habitat fragmentation have measured the extent of isolation of individual habitat fragments, area of the edge and shape of the fragments (Spellerberg and Sawyer, 1999). Such is the importance of analysis of effects of habitat fragmentation that research has been done on analytical methods. For example, some research on forest edges in Australia prompted Salisbury (1993, 1996) to discuss methods of analysis in his *Design for Studying Edge Effects in Forests*.

ROADS AND HABITAT FRAGMENTATION

In general, the effects of habitat fragmentation (by roads) on nature include the following:

- The road subdivides habitats resulting initially in two smaller fragments.
- The 'core' habitat (interior of the habitat) is reduced in area.
- The biota is disturbed and moves away from the road.
- The road acts as a barrier for some animals.
- The road opens up new opportunities for invasion by new species.
- The road creates new edges (ecotones), which favours those species that tend to inhabit edges.
- Road management practices and traffic help some biota to spread into new areas.

The effects of road development on species are very clearly shown conceptually in Fig. 4.3 taken from Sheate and Taylor (1990). They based their work on the assessment of long-term potential environmental impacts on adjacent habitats resulting from motorway developments.

Construction of a single road contributes to fragmentation of habitats because it divides a habitat into two areas. A single road acts as a barrier and the effectiveness of that barrier increases with width of the road and the addition of central concrete or metal barriers. Over time, other roads extend from the one original road culminating in a road network. The network facilitates access and supports further changes in land use and development.

The effects of roads on nature and the impacts of road networks differ. Road densities and their effects have been investigated (see for example in Chap. 2). Theil's (1985) work on road densities and wolf habitats). The most important implication of road networks is habitat fragmentation, which is discussed in a special section below.

Why doesn't the effect of a new road remain within the boundaries of the road itself? The reason it doesn't is because the physical presence of the road

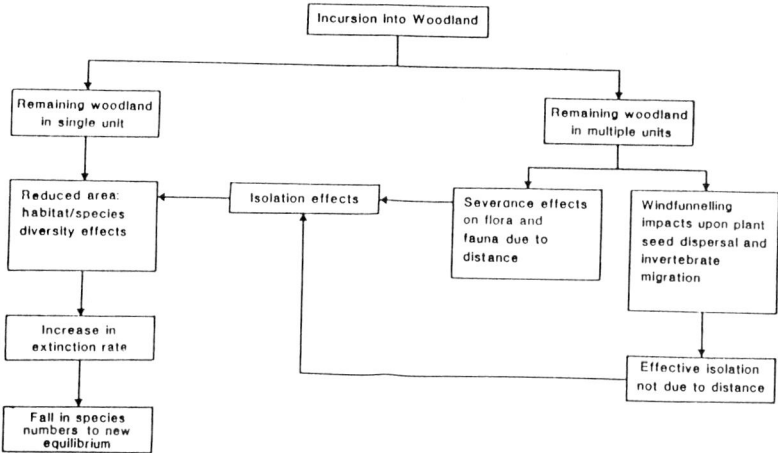

Fig. 4.3. Species effects and biogeographical effects from motorway construction (from Sheate and Taylor, 1990).

changes the physical environment (microclimates, hydrology). It introduces new features and in doing so removes the resources on which some species depend while providing resources for new species. The introduction of any new object to a natural area, be it on land or in the sea, will by its mere physical presence have an effect on nature. It changes the physical structure.

All new roads will inevitably cross or subdivide wildlife habitats. The area taken up by a single road therefore reduces the area of wildlife habitat and roads convert one continuous wildlife habitat into smaller, disjunct areas. A road may act as a barrier for some wildlife and make it difficult for some animals to cross from one side to the other. The new road edges will be colonised by new species and subsequently the road aid in the spread of species new to the area. Road edges may be planted and wildlife subsequently take advantage of these long thin habitats. Some of these linear habitats may provide conduits for movement of wildlife between habitats. Roads may also act as conduits for the dispersal of pests and diseases.

As with habitat fragmentation in general, there have been many studies on roads and habitat fragmentation. As already mentioned, Chomitz and Gray (1996) suggest that for southern Belize, road building in areas with poor agricultural soils and low population densities may be a 'lose-lose' situation, resulting in habitat fragmentation and low economic returns.

Noting that there are different kinds of fragmentation, Harris and Silva-Lopez (1992) introduced the term 'diversive' fragmentation, the result of intrusive roads (as well as rail or power-line swathes) that bisect expansive tracts such that movement of some wildlife is impeded.

The general effects of habitat fragmentation can be broadly grouped into six types, as done by Andrews (1990) after an extensive review of the literature on these effects:

— habitat loss and modification of habitats;
— edge effects and penetration into biological communities;
— isolation of populations (in some cases leading to reduced gene flow);
— traffic causing disturbance to adjacent habitats;
— increased road kills;
— increased human access to nearby habitats;

The incremental effects of roads and consequent fragmentation is illustrated on a small spatial scale by the activities which have taken place in Middlesex Fells, a 400-ha woodland park in metropolitan Boston. This Park is isolated by a barrier of roads. Drayton and Primack (1996) have studied changes in plant species composition since 1894. Exotic species have increased and decreased native species (155 of 422 species have disappeared). They conclude that among the factors contributing to the decline and local extinction of plant species must be included the increased number of trails and roads that dissect the fells into smaller and smaller fragments. This increasing fragmentation reduces the functional interior of the woodland, just as it does in forests. The tracks and roads allow more entry points for exotic and invasive species. For some bird species the fragmented habitat has led to greater pressure from predators. There have also been marked changes in microclimates.

Fragmentation of forest habitats by roads has attracted much attention, leading to the question: Do roads cause more fragmentation than other activities in forests such as clear-cut forestry. This question was addressed in the USA and in one study of forest fragmentation in the Rocky Mountains, Reed et al. (1996) assessed the extent of fragmentation caused by roads on the one hand and that caused by clearcut forestry. They found that roads contributed more to fragmentation than did clear-cuts. Edge habitat created by roads was 1.54-1.98 times the edge habitat created by clear-cuts. The total landscape area affected by clearcuts and roads was 2.5-3.5 times the actual area occupied by these disturbances.

One of the immediate effects of road construction on the environment is to alter the microclimate; this is particularly noticeable for new roads in forested areas. Changes in light levels, temperature levels, wind speed and humidity occur immediately. These changes in microclimate have an instant effect on biota, resulting in plant extinctions and movements of animals away from the edge of the new road. After a while changes in microclimate change the mix of plant species and their local distribution.

Ecological succession (the natural process whereby plants and animals are replaced by others) is also affected. In a study of woodlands in Ohio where Kupfer (1996) studied patterns and determinants of edge vegetation in forest reserves, he concluded that the forest stand microclimate is indeed a significant factor in determining the ecological successional processes at the edges of forest stands.

Changes in microclimate do not end in the area just adjacent to the new road edge but extend into forest stands. Williams-Linera (1990) working on forest edges in tropical premontane wet forests of Panama found changes in microclimate conditions penetrating as much as 15 m into the forests. An edge of the tropical premontane wet forest along the El Llano-Carti road, Panama in the late 1980s is depicted in Fig. 4.4. Several hectares of the primary forest were cut during the dry season and the area burned. With the first rains (April, May), crops were planted for one or two years. The following year the pasture was ready for cattle.

Fig. 4.4. Edge of the tropical premontane wet forest along the El Llano-Carti road, Panama in the late 1980s. This forest edge was 12 years old in 1986/87. Photo Guadalupe Williams-Linera.

Research on fragmented Podocarp-broadleaf forest in North Island, New Zealand by Young and Mitchell (1994) included observations on the microclimate and vegetation edge effects of forest margins created about 100-130 years ago. They found that penetration of edge effects extended to approximately 50 m, regardless of forest size. Some species-specific edge/interior differences could be related to timing of critical life history stages (e.g. germination and early establishment). The authors conclude that these edge processes are probably now a major feature of the overall forest ecology.

Edge effects have now been found to extend farther than 50 m. Working in the Cherokee National Forest in the Southern Appalachian Mountains of the United States, Haskell (2000) calculated that the effects of roads on the abundance of fauna and on depth of leaf litter may persist up to 100 m into the forest. Effects of faunal richness on the other hand extended only up to 15 m.

Effects on Plants

From the edge of a road to the edge of adjacent habitats changes in microclimate occur—light, temperature and humidity—regardless of whether the habitat, be forest, grassland or wetland. These changes determine the ecology of the roadside, especially its dynamic aspects. Roads in wetland areas have devastating implications for the hydrology, including drainage patterns. These effects in turn affect plant communities.

In forests, road edges contain shade-intolerant species. In forest reserves in Ohio, USA for example, Kupfer (1996) attributed composition gradients of woody plant species at forest edges to microclimatic gradients. His and other works indicate that the edge microclimate definitely affects the successional processes on forest edges; that is, it determines what species are present at different stages over time.

Some plant species inhabiting the edges of woodlands and forests may penetrate the interior of woodlands as a result of the opened canopy and changes in microclimates. The biogeography of the woodland changes; consequently species composition and distribution are affected. In the Parque Nacional da Amazonia for example, it was found that species associated with open areas or formations such as *Leptodactylus longirostris* can follow forest clearings (caused by roads) and penetrate forests (Crombie and Heyer, 1983).

Following introduction of a road in a forest, some species may establish on the road edge while others invade the forest interior. Invaders of forest interiors include weed species and considerable concern has been expressed over the widespread phenomenon of invasive weed species in protected areas. Typically, exotic species are more predominant in roadside verges and decline in abundance with increasing distance from the road.

Roads, habitat fragmentation and consequent flow of traffic or people have all contributed to the issue of biological invasions of nature reserves (Usher, 1988). The role of roads (and other linear landscape features such as forest trails and railway lines) in spreading weeds into nature reserves and other protected areas has been well recognised in many countries such as New Zealand (Timmins and Williams, 1990, 1991) and the USA (Tyser and Worley, 1992).

In New Zealand, the Japanese honeysuckle (*Lonicera japonica*) has dispersed along linear habitats beside railway lines up the Kapiti Coast (John Sawyer pers. comm.). In Montana (USA), Tyser and Worley (1992) found 15 alien species adjacent to roads and two Eurasian grasses (*Phleum pratense* and *Poa pratensis*) were particularly common.

Effects on Invertebrates

Compared with the dimensions of most roads, invertebrates are very small indeed. The flat surface of a road must seem a very vast and inhospitable landscape to them. Perhaps not surprisingly therefore, roads have significant effects on the movements of many invertebrates. An extreme example is their effect on the dispersal of snails. Baur and Baur (1990) carried out a number of experiments in central Sweden on this gastropod. They moved snails (*Arianta* sp.) to positions alongside roadside verges and then monitored their movements. Only one snail managed to cross an 8-m wide paved road (low traffic density) and two crossed a 3-m wide unpaved track. Consequent to these experiments the authors concluded that roads with high traffic density could well isolate some snail populations from others.

By far the most systematic and intensive research on road effects on movements and dispersal of invertebrates has been undertaken on beetles and to a lesser extent on butterflies and spiders. Mader's research on movement patterns of beetles around roads in Germany is perhaps the best known of all the pioneering work. He and his fellow researchers (Mader, 1984; Mader et al., 1990) showed that there are indeed barrier effects of roads on carabid beetles. These effects are well illustrated in the classic diagrams shown in Fig. 4.5.

Similar studies (Dennis, 1986) on the orange-tip butterflies (*Anthocharis cardamines*) around motorways in Cheshire, England have also indicated distinct barrier effects (Fig. 4.5). Such effects on spiders is complicated by the fact that although some avoid crossing roads, most lycosid spiders disperse aerially when immature (they float across using a thin thread).

Effects on Amphibians and Reptiles

Throughout the world, there is a worrying decline in amphibians (frogs, toads, newts and salamanders). One of the most important factors contributing to this decline in amphibians, at least in industrialised area, is habitat fragmentation and the fact that roads interfere with natural seasonal migrations of some amphibians.

Some species of amphibians (as well as some snake and mammal species) have well-established travel routes. Notable examples are the routes taken by some toad species between wintering sites and aquatic habitats used in summer. Other examples are the daily routes used by some mammals as they move from day time refuges to night-time feeding areas. On occasions, new roads cut across routes that may have been used for decades. A road cutting through a well-established route can be devastating for wildlife, resulting sometimes in high mortality as the animals attempt to cross the road in the line of the old route used for migration or daily movements.

One of the most detailed studies on the effects of habitat fragmentation and road density on frogs was undertaken in the Netherlands. Vos (1997, 1999) and Vos and Chardon (1998) researched the effects of habitat fragmentation on the distribution pattern of the moor frog (*Rana arvalis*) and tree frog (*Hyla arborea*). Habitat fragmentation partly explained the distribution pattern and it appears that road density may have a negative effect on the probability of inhabiting a moorland pond. Overall, Vos and Chardon believe that the negative effects of roads on fragmentation of amphibian

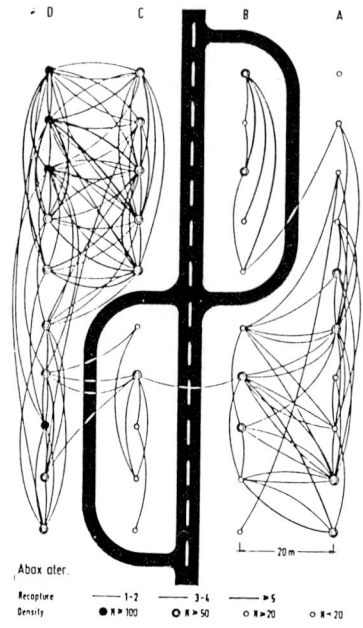
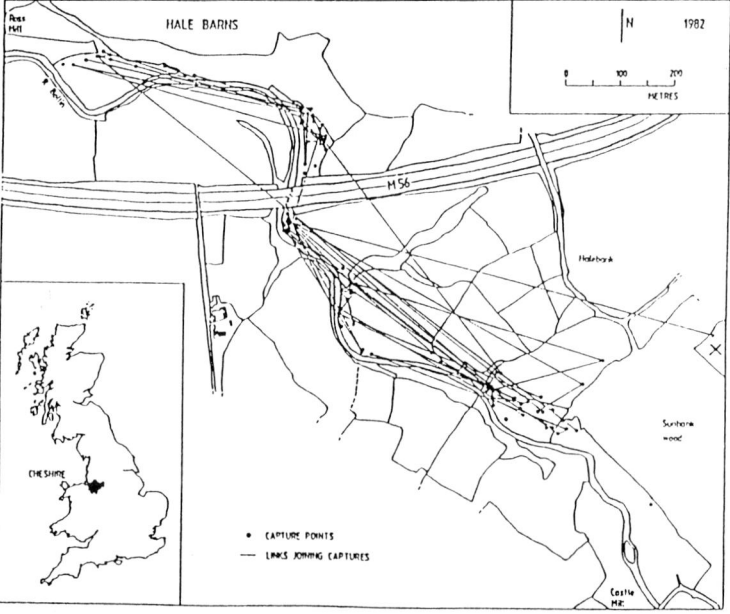

Fig. 4.5. Top: Diagram showing the barrier effects of roads on beetles (*Abax ater*). A-D are trap rows, circles live traps and curved lines represent movement of marked beetles between captures (from Mader, 1984). **Bottom:** Movement of butterflies in relation to roads. This shows capture points and lines joining successive recaptures for male orange tips (*Anthocharis cardamines*) in the Bollen Valley during 1982 (from Dennis, 1986).

habitats is often underestimated. They conclude from their research that roads increase habitat fragmentation, especially isolation of habitat patches, and therefore negatively affect the survival probability of amphibian populations.

By way of contrast, the negative effects of roads on amphibians in Europe, in Australia, roads and vehicle tracks are of benefit to the introduced cane toads (*Bufo marinus*). These toads, introduced in 1935, use roads and tracks for dispersal, especially in forested habitats and therefore roads appear to have assisted them in extending their range (Seabrook and Dettman, 1996).

Effects on Birds

The effects of habitat fragmentation on birds have been widely studied in forest locations and are similar to those found for other groups of animals. Forest habitat fragmentation results in decreased levels of resources for the avifauna. For example, forest fragmentation may lead to reduced levels of standing dead wood and that in turn may have implications for the species composition of tree-cavity nesting birds.

The widespread fragmentation of what formerly were largely unbroken forests in eastern North America (together with loss of habitats) is believed to be the primary contributor to decline in forest bird species (literature reviewed in Rich et al., 1994). This decline comes about in part because habitat fragmentation exposes the birds to nest predation and parasitism of the broods, resulting in low reproductive success.

In southern New Jersey (USA), Rich et al. (1994) examined the effects of three types of ubiquitous, narrow, forest-dividing corridors on the relative abundance and community composition of forest-nesting birds. The corridors were unpaved roads (8 m wide), paved roads (16 m wide) and power-line swathes (23 m wide). They found that forest-interior species had significantly reduced relative abundances on edge transects along 16 m and 23 m corridors compared with 8 m corridors. They also found that even narrow forest-dividing corridors (8 m) do affect the distribution and abundance of birds in ways that are associated typically with the effects of forest fragmentation. Most importantly, they found that even small width corridors in forests may function as 'ecological traps'. That is, in highly fragmented forests, some bird species are attracted to nest near forest edges, where predation rates are highest (referred to as 'ecological traps' by Gates and Gysel, 1978).

Effects on Mammals

Research on the effects of habitat fragmentation (brought about mainly by roads) on mammals can be broadly divided into effects on large or on small mammals. Elephants, bears, badgers, koala bears, mountain lions and deer are just some of the large mammals that feature in studies on the effects of roads and habitat fragmentation. One indirect effect of roads on large mammals is increased vulnerability to hunters (a major source of bear mortality).

In North America, the impacts created by highways are considered a severe conservation issue for bears, wolves, the lynx and the eastern cougar (e.g. Ruediger, 1996). Measures to address fragmentation effects are being implemented (Servheen et al., 1998). For example, the effects of roads on the behaviour, habitat use and demography of grizzly bears in the Rocky Mountains has been studied by McLellan and Shackleton (1988). They found that these bears use habitats near roads less and that this reduction in use accounts for over 8% habitat loss. Roads affect movements of not only grizzlies, but also black bears. In western North Carolina, a study of black bears by Brody and Pelton (1989) found that these bears almost never crossed an interstate highway. Of the roads the bears did cross, those of low traffic density were crossed more frequently. Like bears, mountain lions (*Felis concolor*) are less likely to cross paved roads than improved dirt roads and tracks (van Dyke et al., 1986).

In Europe, highways seem to be a barrier rarely crossed by bears. In 1996, Kaczensky et al. assessed the impact of highways on brown bears in Slovenia. They found that the more fragmented the landscape, the smaller the bear population. They therefore emphasise the need for brown-bear crossings.

Fragmentation of habitats may have implications for dispersal by feral predators but little research seems to have been done in this area. May and Norton (1996) noted in Australia that the extent to which roads influence the distribution and abundance of species such as foxes, cats and dingoes, and the consequences for native fauna are poorly known.

Studies on habitat fragmentation and effects of roads on small mammals have shown that mobility can be affected. In Kansas (USA) for example, small road clearances less than 3 m were proven to affect movements of voles and rats (Swihart and Slade, 1984). In Australia, Mansergh and Scotts (1989) showed that the social organisation and survival rate of the mountain pigmy-possum (*Burramys parvus*) has been disrupted due to habitat fragmentation by roads and other developments within a ski resort. The authors suggest underroad corridors to aid dispersal.

Considerable research is now available on various species of small mammals, showing that roads constitute barriers although some are not insurmountable for some species (see for example Adams and Geis, 1983; Bakowski and Kozakiewicz, 1988; Oxley et al., 1974; Richardson et al., 1997; Swihart and Slade, 1984).

Despite the extensive number of reports on the mortality and barrier effects of roads on small mammals, very little research has been done on the consequences for the small mammal populations.

ROADS AIDING DISPERSAL OF PLANTS AND ANIMALS

Roads and traffic have long been known to aid the dispersal of some species viz., fungi, plants and mammals.

Traffic can spread pathogenic fungi and apparently such pathogens have infected forests in southern Australia (Weste, 1977).

Williams and Buxton (1995) surveyed vegetation invaded by two species of *Passiflora* in South Island, New Zealand and draw attention to records from road cuttings. Research on weeds in New Zealand forest and the implications of reserve design for weed control has been reported by Timmins and Williams (1990, 1991). Although not specific to New Zealand, Crawley (1989) from Auckland University has reviewed the chance and timing of biological invasions, a topic very relevant to the ecology of dispersal of introduced species.

In a study of stoats inhabiting New Zealand beech forests, Murphy and Dowding (1994) found that a road through the study area affected their behaviour. Females avoided the road while males preferred it. More recently, King et al. (1996) have reported studies on small mammals in relation to habitat in Pureora Forest park, central North Island. They note that roads introduce a linear community of small mammals into forests.

In Illinois (USA), Warner (1985) studied the movements of free-ranging domestic cats and found that they made disproportionately high use of farmsteads, roadsides and field boundaries.

The role of motor vehicles in the dispersal of seeds or plants has been known since at least the early 1930s (Clifford, 1959; Amor and Stevens, 1975; Schmidt, 1989; Cowie and Werner, 1993; Timmins and Williams, 1990, 1991; Lonsdale and Lane, 1994). Clifford's work was based on mud samples taken from vehicles in Nigeria in which 38 plant species later germinated. In Germany, a study of seeds found in the mud on a car driven 15,000 km in a single journey revealed 124 species that yielded 3926 seedlings (Schmidt, 1989).

Lonsdale and Lanes (1994) in their work on weed seeds transported by tourist vehicles in Kakadu National park, Australia suggest that rather than attempting to prevent this form of seed movement, resources could more profitably be spent on detecting and eradicating weed infestations. This conclusion is based on acceptance of the fact that the road has been built and is operational. Furthermore, ongoing costs have to be considered.

Also from Australia is an excellent analysis of the mechanism of dispersal given by Wace (1977; *'Assessment of dispersal of plant species—the carborne flora in Canberra'*). He assessed the size and quality of carborne flora in Canberra for 2.5 years, during which period 18,566 seedlings were germinated from carwash sludge and side-scrapings and 259 species (or taxa of a higher rank) recorded.

Dispersal of weeds and alien flora via roads and road traffic (and by vehicles and humans during road construction) has received attention in New Zealand. Timmins and Williams (1990, 1991) for example looked at the accidental spread of weeds through reserves and noted that reduced roading was the best means of addressing this problem.

There is no denying that the huge volume of road transport poses considerable risk for the spread of weeds. Traditional mechanised commercial car-washing is known to be merely a cosmetic operation leaving an abundance of weed seeds in the vehicleborne biota (Wace, 1977). More effective cleaning methods need to be developed, especially for imported vehicles.

Not withstanding the studies mentioned here, the extent of car-borne flora has not been well researched to date. Indeed, in their review of plant dispersal and the role of man, Hodkinson and Thompson (1997) note the paucity of dispersal of literature concerning seed dispersal by vehicles in spite of the fact that the present-day volume of traffic ensures a far higher level of disperse than ever before. There is concern, however, over vehicular transport of alien and invasive weed species into protected areas, especially in the USA. Weeds are indeed transported into national parks by vehicles but it is believed that rather than attempting prevention of this movement, it is better to detect and eradicate existing weed infestations.

That roads and road traffic can facilitate dispersal of species, including pest species, cannot be disputed. Research into the ecology of dispersal of alien species and the ecology of invasions has taken on considerable importance and concern. A scientific journal *Biological Invasions* is devoted to this broad field. In Britain, for example, Usher (1988) reviewed biological invasions and noted that tourism poses dangers for reserves since a positive correlation was established between visitation rate and number of introduced species.

The spread of alien and invasive species into nature reserves has evoked particular concern. Brothers and Spingarn (1992) working in central Indiana, USA, for example, have drawn attention to the possibility of forest fragmentation (which can be caused by road construction) giving rise to invasions by alien species for at least two reasons. First, fragmentation increases the ratio of non-forest to forest and of forest edge to interior. Second, microenvironmental changes at forest edges may provide points of entry for alien species. Elsewhere, in New Zealand for example, Timmins and Williams (1991) have noted that the most important factors, among other influencing the number of problem weeds in nature reserves are distance from roads and railway lines. Other significant studies and reviews of invasions of nature reserves by introduced species have been done by MacDonald and Frame (1988) and MacDonald et al. (1989).

ROADS AND ROADSIDE VERGES AS LINEAR HABITATS

Much has been made of the observation that road verges and highway right-of-ways do support nature. For example, some bird species obtain grit from roads, electricity power lines provide perches for birds and roadside verges may provide habitats for certain plants and animals. Road verges as potential wildlife habitats have, in some countries, been advocated as one of the benefits of roads for wildlife. However, when roads are established a net loss in habitat does occur and the road verge is small compensation for this loss. On the other hand, in countries where natural habitats or seminatural habitats are few, road verges play a valuable, even if small, role by contributing to or augmenting the networks of nature reserves.

Road verges in some instances are, ecologically speaking, more than a boundary delimiting a road. In ecological communities, an edge is an important structural feature i.e., a boundary between two kinds of biological habitats or between two different kinds of land use. In ecology, this edge is called an ecotone (see Appendix I for definition). The creation of these ecotones may benefit a number of species. For example, Evans and Gates (1997) working in Maryland, USA found that cowbirds (*Molothrus ater*) inhabit forest edge habitats or ecotones created by roads (cowbirds are considered 'weed birds'). However, such benefits are small compensation for loss of habitats.

In many countries, interest in linear habitats (such as those provided by road verges, edges of railway lines, and edges of rivers and streams) has grown markedly. In America, for example, Smith and Hellmund (1993) have reviewed what they term 'greenways' (a linear open space or a corridor composed of natural vegetation) and provide an overview of greenway ecology and habitat management. They quite rightly say that while large patches of natural vegetation remain the top conservation priority, greenways or linear habitats can provide links or connectivity between patches.

What is planted on a strip of land along a road and how it is managed can make all the difference to the composition of wildlife. For example, Roach and Kirkpatrick (1985) reviewed the wildlife use of roadside woody plantings in Indiana, USA, and suggested (as alternatives to predominantly monocultures of closely mowed grass), the planting of trees and shrubs along certain four-lane highways. They found that such planted areas were used by a greater number of bird species in much greater abundance than grassed areas.

The UK has also shown considerable interest in the contribution of roadside verges to conservation. Way reviewed roadside verges and conservation in Britain bank in 1977. A detailed analysis of one county's roadside verges was undertaken for Cheshire (Cheshire Ecological Services, 1995 (see also Chap. 3)). This report details survey methods, analyses roadside verge habitats and makes recommendations for management and monitoring.

WILDLIFE CORRIDORS?

Roadside verges and highway right-of-ways are linear habitats. They also function as conduits for dispersal. This dispersal is often only in the sense of dispersion and not in the sense of a linear habitat acting as a corridor for movement of genes between populations in habitat patches. There are many projects on wildlife corridors ranging in scale from a few km to many thousands of ha. In Australia for example, there is the 'Big to Little Desert

Biolink Corridor'. Located in Victoria, it offers the possibility of creating the largest road corridor network within the island continent (Straker, 1998).

The corridor concept is very simple. Despite the amount of work being done and the establishment of such corridors and regional networks, very little has been said as to whether or not they work. This is not to say that corridors and networks should not be supported; on the contrary! But emphasis has been given to aims and design while research on effectiveness lags behind.

Much information is available on the positive role roads play in the dispersal of species, focusing in particular on the dispersal of exotic species (see for example Parendes and Jones, 2000). But the effect of such corridors can also be negative, luring some animals (such as bison) away from the safe haven of nature reserves.

While there is evidence that some species of plants and animals use linear habitats as corridors for movement between habitat patches, research on this topic is rather scant. The possible corridor function has been widely popularised and the term wildlife corridor is commonly used. True, many linear landscape features such as roadside verges may provide linear habitats, but evidence actually supporting use of roadside verges by wildlife as conduits for dispersal is still minimal (Spellerberg and Gaywood, 1993). Interest in this topic is rapidly growing, however. Corridors for wildlife can be considered in two ways: potential landscape linear features, usually associated with human impacts on the environment and as naturally occurring migration corridors used by wildlife. Studies of the naturally occurring corridors or migration routes within the context of biogeographical research are extremely important in addressing the effects of the transportation infrastructure on wildlife habitats and the extensive habitat fragmentation that has taken place.

ADDRESSING THE PROBLEMS CAUSED BY ROADS AND FRAGMENTATION

In general what can be done to avoid, remedy and mitigate the effects of habitat fragmentation caused by roads? The simple answer is not to build the road in the first place. That answer may not be quite as simplistic as it appears, however. There may be alternative options—a tunnel or a bridge, perhaps. Such thinking often leads to the debate 'what price wildlife'?

Carola Stauch, previously at the University of Stuttgart, has made use of Geographical Information Systems (GIS) in spatial decision support systems as applied to the effects of fragmentation by roads (Stauch, 2000). It is possible for example to calculate the extent of sensitive areas affected by each road section and then to rank these sections. Criteria at one level include fragmentation and at another traffic characteristics, road characteristics and adjacent habitats (Stauch, 1998). One example given is that while it may not be possible to compensate for the effects on sensitive wetland areas by providing a protected area, quite possibly slower traffic speed could compensate for the effects of traffic on birds.

The loss and degradation of habitats can to a small extent be mitigated by ecological restoration of roadside verges or, even better, roadside nature reserves or preserves. The area of habitat taken up by the road may be offset by restoration of at least the same area of degraded habitat elsewhere (see reference to conservation banking in Chap. 7).

In general, habitat fragmentation by proposed roads can be avoided by the following:

— weighing the long-term costs and benefits of establishing a road;
— considering alternative transport routes;
— planning the route so that it avoids naturally occurring vegetation and significant habitats;
— controlling traffic density and traffic speed in areas where wildlife is likely to be disturbed;
— restoration of habitats;
— connectivity mechanisms such as tunnels (underpasses) and wildlife bridges.

In cities, little space is available for offsetting the land taken up by roads apart from planning for the introduction of or expansion of parks and nature reserves. One innovative project, which addresses habitat loss in urban areas, is the proposal to use traffic islands and dividing strips for indigenous plant species and even endangered plant species (Fig. 4.6). The roads that surround the plants on the traffic islands may afford protection for the plants from vertebrate grazing animals. However, the dangers from traffic may continue to exist.

Sharing right-of-ways has been advanced as a means of mitigating the fragmentation impacts of land transportation corridors. Feitelson and Papay (1999) investigated where, when and under what circumstances right-of-ways in Israel have been actually shared. They found them to be shared only to a limited extent and mainly within core areas.

For many years now, tunnels or fauna underpasses (passages) have been designed and installed under roads for wildlife (in particular mammals and amphibians). They have been assessed for their effectiveness in certain instances (see Clevenger and Waltho, 2000 for example). In the Netherlands, faunal passages have proven suitable for small mammals and the Directorate-General of Public works and Water Management have undertaken research on many designs. The entrance to pipes or culverts can usefully be modified to serve as faunal passages (Fig. 4.7). Footprint tubes have helped in assessing the effectiveness of these faunal passages; an example is depicted in Fig. 4.7. 'Drainage culverts have only recently been investigated as habitat linkages (Clevenger et al., 2001).

If the wildlife can't go underneath then perhaps overpasses are the answer. The introduction of wildlife bridges (green bridges, wildlife overpasses,

Fig. 4.6. Rare plants on central reservations in Wellington City, New Zealand. Included are *Aciphylla diefferbachii*, *A. squarrosa* and *Carex testacea*. Photo Tony Silbery.

ecoducts) in Europe and North America has attracted much attention and discussion. Such bridges are not a new idea; they have been constructed in France since 1981 and the design used at that time has been adopted elsewhere. In Canada, there are two wildlife bridges (and twelve underpasses) on the Trans-Canada Highway. The USA seems to be lagging behind on the issue of wildlife bridges. In the Netherlands, two wildlife bridges (ecoducts)

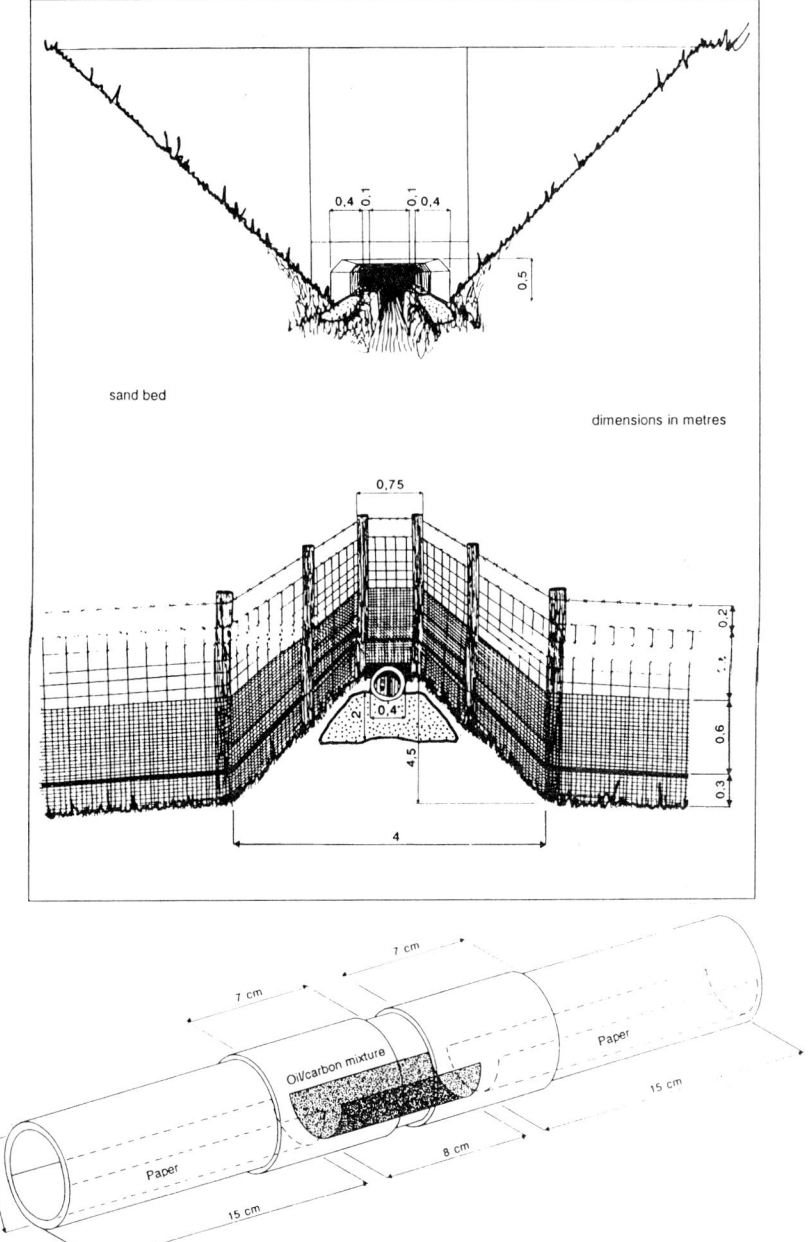

Fig. 4.7. Top: Two examples of designs for entrances to culverts.

Bottom: Example of a footprint tube. Reproduced with permission from Nieuwenhuizen and van Apeldoorn (1995), *Mammal use of faunal passages on national road A1 at Oldenzal*. Ministry of Transport, Public Works and Water Management, The Netherlands.

were built in 1988. Wildlife bridges were initially designed for red deer (*Cervus elaphus*).

The effectiveness of wildlife bridges (and that of underpasses) has been the focus of recent research (see for example Foster and Humphrey, 1992; Lotz et al., 1996; Roof and Wooding, 1996). In some cases it has been shown that wildlife bridges for large mammals do fulfil their purpose (see for example Keller and Pfister, 1997). On the other hand, Janssen et al., (1997) have demonstrated that many underpasses built for badgers do not meet the criteria for good technical installation.

The cost of building wildlife bridges for motorways has concomitantly attracted attention, focusing on costs of bridges in relation to total outlay for motorway construction. Some say the costs are relatively small while others argue that they are significant. Can the costs be justified? This is a difficult question and broadly speaking a misleading one. The costs of projects such as motorways must surely take into consideration the costs of mitigating their effects on nature and the environment.

Who pays for the bridges seems also to be a misleading question. Bridge costs should be seen as part of the motorway costs and should be funded as part of the whole project.

The biogeographical and ecological issues of green bridges have only just recently come up for assessment. For example, Jackson and Griffin (1998) advocate analysis first at the landscape level to ensure realisation of landscape continuity and metapopulation dynamics within 'connectivity zones'. This approach includes an evaluation of landscape features to determine the most valuable habitat for wildlife and wildlife movement.

Biogeographical and ecological studies are certainly required to determine where the main travel corridors lie for certain species. That information is essential in determining proper location of wildlife bridges. This all seems commonsensical but common sense does not always prevail.

In Austria for example, there was a call for some years to establish wildlife bridges over the motorway between Vienna and Budapest. Agreement for such bridges came at a very late stage in the motorway project. They were indeed built (Fig. 4.8) but not at places that would facilitate movement of large mammals. In other words, they were constructed in the wrong locations (Friedrich Voelk, pers. comm.).

Six wildlife bridges on the Vienna-Budapest Motorway (each 100 metres in width) were spaced every km. The aerial photograph in Fig. 4.8 clearly shows that the bridges are surrounded by agricultural land.

To resolve the problem of non-utilisation of these bridges by wildlife, the width of the bridges, the distance between them, the nature of the habitat on the bridges per se as well as the nature of the surrounding habitat are all important factors that need to be addressed. Ecological and biogeographical studies can help to provide some of answers.

Friedrich Voelk and others at Boku University in Vienna have undertaken an extensive programme to examine the issues of habitat fragmentation

Fig. 4.8. Wildlife bridges in Austria. These six bridges spaced 1 km apart and 100 m wide span the motorway between Vienna and Budapest (aerial photo, Friedrich Voelk).

caused by motorways. They have assessed the effectiveness of motorway passages in facilitating movement of large mammals (Fig. 4.9a) and as a basis for determining the locations and design of wildlife bridges, analysed the habitat connectivity for big game species (Fig 4.9b).

There are many guidelines and policies for wildlife bridges and passageways but a notable lack of standards. Friedrich Voelk and Irene Glitzner therefore developed a checklist and questionnaire for collecting information about bridges to help planners identify minimum standards for width of bridges and the number needed (Box 4.1). Full details can be found in Voelk *et al.*, 2000, Voelk, 2001).

The extent of development of road networks and motorways throughout Europe shows no sign of slowing down and calls for the construction of roads continue to increase. The demand for improved road networks is due in part to the growing trade between East and West (Box 4.2). Consequently, the link between infrastructure and extent of habitat fragmentation has been the basis for much discussion. For example, a major international conference was held in September 1995 in The Hague with 315 participants (Canters et al., 1997). In October 1999, the Conference 'Fauna and traffic: traffic systems and fauna networks, necessity for a new approach' was held in Lausanne (Dumont et al., 1999).

The scale of habitat fragmentation throughout Europe and the effects of new roads on this scale require both a national and an international approach to dealing with the problems. In the Netherlands, the National Ecological Network has been established to halt the continuing process of habitat fragmentation. Also in the Netherlands, the spatial concept of traffic-calmed areas in rural areas has been suggested (Jaarsma and Langevelde, 1997). The approach suggests a fundamental change in rural transportation planning, from following actual traffic flows to regulation of traffic flows. According to the authors, this could result in the following:

— diffuse volumes of traffic concentrated on fewer roads,
— volumes of traffic and speed within regions decreased,
— decreased habitat fragmentation.

A European Network of Experts and Institutions represented by the Infra Eco Network Europe (IENE) is involved in habitat fragmentation issues which have been brought about by road networks and a general growth in the transportation infrastructure. IENE promotes co-operation and exchange of expertise. The overall goal is safe and sustainable transportation infrastructure while at the same time conserving wildlife. IENE can be accessed on http://www.iene.vv.se.

The European Commission Directorate for General Transport together with COST (Co-operation in the Field of Scientific and Technical Research) have supported a project (COST 341) to examine habitat fragmentation due to transportation infrastructure. The main objective is to promote a safe and sustainable pan-European transport infrastructure through recommending

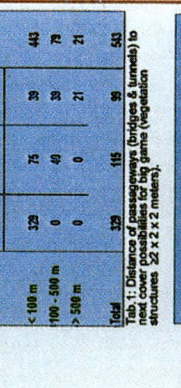

Fig. 4.9a. Posters on road infrastructures and wildlife (Friedrich Voelk, Boku University, Vienna).

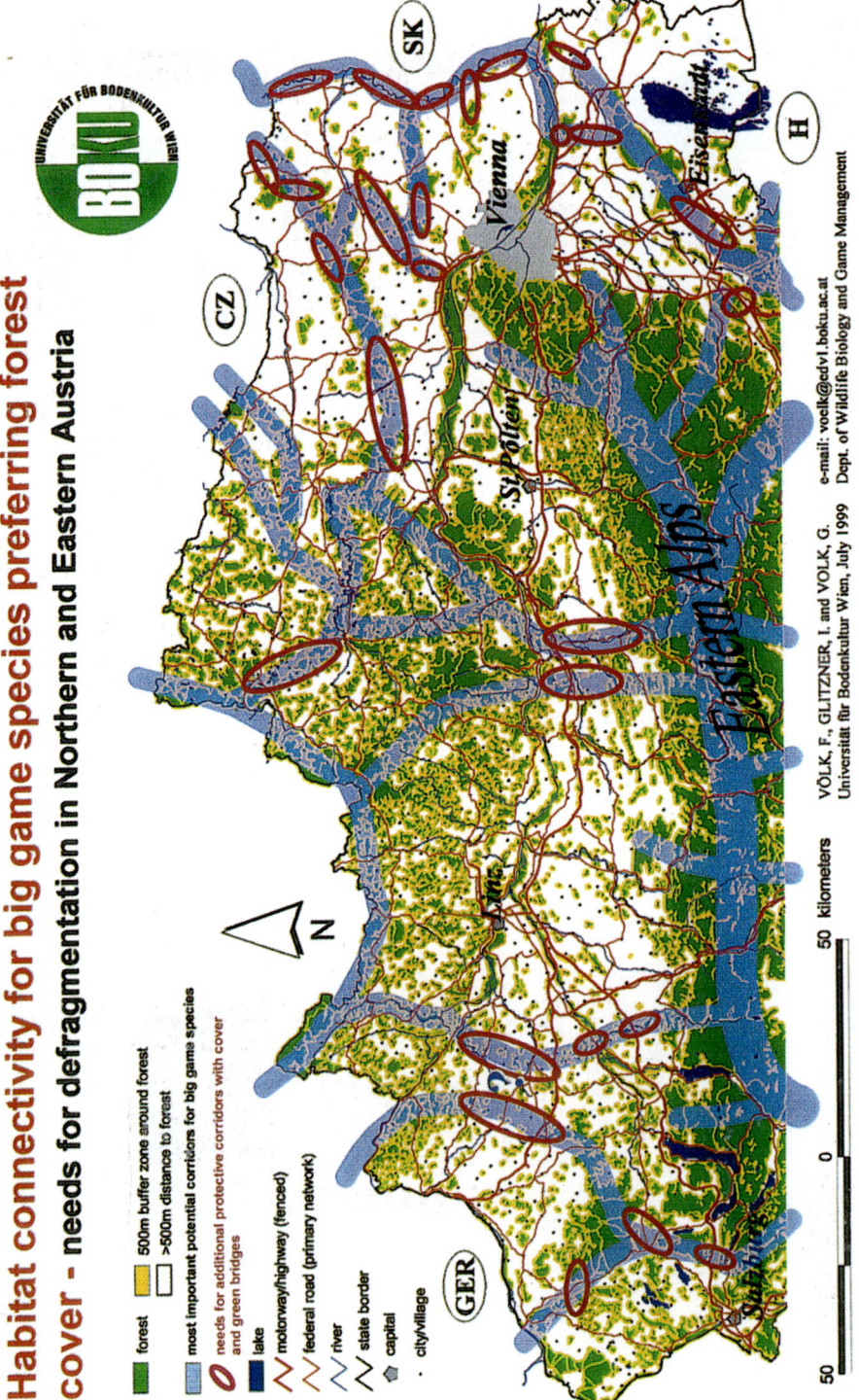

Fig. 4.9b. Posters on road infrastructures and wildlife (Friedrich Voelk, Boku University, Vienna).

Box 4.1:

*Designed by Friedrich Voelk, Boku University, Vienna and translated from the German by Jean-Paul Thull.

Form 1: 'Preparation' with assistance of a map scaled 1: 50,000

DATE		PERSON IN CHARGE		at km		
NAME & NUMBER OF	Motorway		State Highway		Provincial Highway	

1. Game is crossing over the road
- ☐ Underpass, road bridge over valley — bridge length .... m
- ☐ Tunnel — length .... m
- ☐ Overpass (combination of driveway or footpath possible) — length .... m

Game is crossing over the road
- ■ Height above Sea-level: .... m
- ■ Name of construction, if known:

2. Type of 'road' under the construction (or above the tunnel)
☐ Railway ☐ Motorway ☐ State Highway ☐ Provincial Highway ☐ Stream ☐ River
☐ Shingle Road ☐ Passage for farm vehicle ☐ Footpath ☐ Tramping Path

3. Landscape elements in the vicinity of the game crossing facilities (horizontal distance to the construction axis < 500 m (for underpasses, bridges, tunnel)
☐ Meadow/Paddock ☐ Vineyard ☐ Moor ☐ Waters ☐ Single Shrubs

4. Closest Forest area (on map 1:50,000)

closest forest area:	☐ < 100m	☐ 100 < 500m	☐ 500 < 1000m	☐ > 1000m
second closest forest area:	☐ < 100m	☐ 100 < 500m	☐ 500 < 1000m	☐ > 1000m

5. Closest features within the following distances

1-2 buildings:	☐ < 100m	☐ 100 < 500m	☐ 500 < 100m	☐ > 1000m
3-10 buildings:	☐ < 100m	☐ 100 < 500m	☐ 500 < 100m	☐ > 1000m
> 10 buildings:	☐ < 100m	☐ 100< 500m	☐ 500 < 100m	☐ > 1000m
river:	☐ < 100m	☐ 100< 500m	☐ 500 < 100m	☐ > 1000m
railway:	☐ < 100m	☐ 100< 500m	☐ 500 < 100m	☐ > 1000m
road 1 order (state highw.):	☐ < 100m	☐ 100< 500m	☐ 500 < 100m	☐ > 1000m
road 2 order:	☐ < 100m	☐ 100< 500m	☐ 500 < 100m	☐ > 1000m
road 3 order (rural road):	☐ < 100m	☐ 100< 500m	☐ 500 < 100m	☐ > 1000m
footpath (not marked):	☐ < 100m	☐ 100< 500m	☐ 500 < 100m	☐ > 1000m
tramping path (marked):	☐ < 100m	☐ 100< 500m	☐ 500 < 100m	☐ > 1000m

6. Game ecology
Pros: +
 +
 +

Game ecology
Cons: –
 –
 –

> **Box 4.2:**
>
> **Pave the Fields, Dam the Rivers**
>
> During the Communist era, several highly-damaging, large-scale industrial projects were built in Poland. The Communist legacy includes the Lenin Steel Works (renamed the Tadeusz Sendzimir Steel Works) near Krakow, which is not only a notorious polluter, but was also built on top of some of the best farmland in Poland.
>
> History looks set to repeat itself in the form of two new large-scale infrastructure projects: a 2600 kilometre superhighway network and a series of cascade dams on the Vistula river, Poland's main river artery. Both projects have long been on the drawing board, but were never realised because of their high costs and other priorities. Today, with Western finance, the projects may soon become a reality.
>
> The demand for an improved road network in Poland has increased in recent years, owing in large part to the growing trade between East and West. With only 300 kilometres of existing highway, Poland has a reputation of being a giant speed bump for trucks travelling along the major transport axes of Europe.
>
> With preliminary costs already hovering above $10 billion, the Polish Ministry of Transport predicts an economic boom and promises 200,000 new jobs. Yet as Pawel Rozynski, a reporter with *Gazeta Wyborcza*, indicates, "roads are constructed with modern equipment, not spades." Not only are new jobs doubtful, but the impacts on other sectors such as agriculture (through farm consolidation) and rail transport are likely to result in reduced employment.
>
> Funds for the highway are expected to come from private investors. The European Investment Bank has agreed to lend Poland $1.2 billion over three years to finance construction. Last year, Minister Boguslaw Liberadzki announced that 'fifteen large US companies are interested in the construction and operation of motorways in Poland'. More recently, the US trade development agency granted $1 million to Poland for preliminary work on motorway development.
>
> Between 10 and 20 Polish consortia are expected to bid for the highway construction. However, with little capital of their own, it is unlikely that banks will loan them the millions of dollars needed for construction. Western consortia are thus more likely to be awarded the contracts.
>
> Although interest in construction is high, investors are demanding financial guarantees from the government. The Polish Ministry of Finance has agreed to guarantee up to 50 per cent of the project's value. The risk will thus be handed over to the average Pole, who, if the project fails, will be burdened with a massive debt.
>
> Tollbooths built at regular intervals along the highways are expected to generate most of the income to pay back lenders. But generating sufficient
>
> Contd.

> **Box 4.2 Contd.**
>
> revenue will require an enormous growth in traffic. To meet the interest payments alone on the loans, the number of vehicle miles travelled in Poland will have to triple.
>
> Poland still relies to a large extent on railways and small, fuel-efficient vehicles. Local and regional improvements are needed rather than large arterial thoroughfares. As writer Alina Ploszaj indicates, "Every Polish enthusiast of motorization would prefer good, safe, roads in their own neighbourhood to good, safe roads intended for German and Russian transit".
>
> The plan for a series of seven cascade dams on the Vistula River follows a similar logic to that of the highways. Revenues from the generation of hydroelectric power are expected to pay back the $3 billion needed for the project. At the same time, the dams will ruin one of Poland's most prized natural resources.
>
> The Vistula's 600-kilometre course, often bordered by lush marshy forests, is valued as a prime habitat for birds. Flow controls would irrevocably alter the ecology of the river and quickly destroy what is recognised as a vital European bird sanctuary. For Poland, it would also mean the ruin of a powerful cultural metaphor, as the free-flowing Vistula has long been a symbol of the country's hope for independence.
>
> *The Ecologist*, 26, 1996

measures and planning procedures with the aim of conserving biological diversity and reducing wildlife road casualties. A report is currently under preparation. Information on COST Transport can be accessed through the CORDIS server at http://www.cordis.lu/cost-transport/home.html.

5

Physical and Chemical Effects of Roads and Traffic

INTRODUCTION

Construction of some roads may result in erosion, slips and transport of sediment to aquatic habitats. From an engineering perspective this can be the cause of long-lasting impact on the natural environment. Fortunately, there are now many case studies from which lessons can be drawn.

This chapter is divided into brief sections on the basis of types of physical and chemical disturbance. However, there are both primary and secondary effects arising from these disturbances and of course there are synergistic effects, some of which may be very complex. For example, road pollutants may cause physiological stress in some plants and make them more susceptible to pest attacks (see for example Braun and Fluckiger's (1984a, b) work on aphid infestations of roadside trees in Switzerland).

LONG-DISTANCE AND GENERAL EFFECTS

Vehicles on roads are the cause of both physical and chemical effects on the environment, some of which have been linked to impacts on biota and ecosystems. Physical and chemical impacts on biota include noise, light, sand, dust and other particulates, metals such as Pb, Cd, Ni and Zn, gases such as CO and NO_x. The range of pollutants (from spillages, vehicles, road surfaces) in water runoff is so large (see for example Gjessing et al., 1984) that this has resulted in very different levels for the study for the different pollutants; consequently some pollutants have been subjected to very comprehensive studies while others have been barely studied if at all.

Knowing the effect of physical and chemical disturbances on biota and biological communities is one thing, but equally important has been research on the extent to which emissions and disturbances extend from a road. In general, it appears that the effects from some physical and chemical agents can be detected some hundred of metres from a road (Schonewald-Cox and Buechner, 1992) and it has been concluded by some authors that impact statements concerning roads which disregard long-distance effects on fauna should be rejected (Van der Zande et al., 1980). Some examples are given in Table 5.1.

Table 5.1: Examples of the extent to which road-generated effects penetrate adjacent habitat (from Sconewald-Cox and Buechner, 1992) with permission of Kluwer Acdemic Publishers.

Road Effect	Distance	Reference and Location
High levels of heavy metals:	Up to 48 m	Scanlon 1987; Virginia, USA
In soils near highways	30 m	Warren and Birch, 1987;
In plants	40 m	England
	120 m	Foner 1987; Israel
		Deroanne-Bauvin and
		Impens, 1987;
In animals	up to 48 m	Netherlands
		Scanlon, 1987; Virginia, USA.
Sediment reaching areas needing protection	9 m-89 m	Hynson et al., 1982; USA.
Increase in edge-species Component of bird community	100 m	Ferris, 1979; Maine, USA.
Plant damage from de-icing salts	120 m	Simini and Leone, 1986; USA
Decrease in abundance of elk	125 m-500 m	Witmar and de Calseta, 1985; Oregon, USA
Altered activity budget of caribou	300 m-600 m	Murphy and Curatolo, 1987; Alaska
Altered abundance of some bird species	400 m	Ferris, 1979; Maine, USA
	200 m-1200 m	Van der Zande et al., 1980; Netherlands

Can any generalization be made about the pollution effects of roads? In one review of toxic substances in flowing water, Hellawell (1988) described the wide range of potential pollutants and noted that very few generalisations can be made about their effects on biota. Each species tends to respond to different pollutants in different ways and even different stages in the life history may have very different responses.

EFFECTS OF OFF-ROAD VEHICLES

Vehicles are not of course restricted to using roads. Many vehicles can be driven off roads and indeed off-road activities in vehicles is widespread and becoming a popular recreational activity. In the USA in particular, off-road vehicles use is increasing dramatically. Concern about the impacts of off-road vehicles on nature has long been registered. In some instances the tracks left by off-road vehicles can remain evident for decades. Intensive use of off-road vehicles may lead to erosion, degradation of habitats as well as loss of wildlife. There are also conflicts with amenity, cultural and land uses. This topic appears to be poorly researched apart from the effects of off-road vehicles in some biogeographical regions (e.g. Webb & Wilshire, a book published in 1983 on the effects of off-road vehicles in arid regions) and falls outside the scope of this book.

CONSTRUCTION OF ROADS AND ROAD DESIGN

Roads, whether meant to be temporary or permanent, have sociocultural effects, economic effects and physical effects. The overall physical effects are clearly shown in Sheate and Taylor's conceptual diagram (Fig. 5.1). A common physical effect is ground compaction, which can have serious implications for drainage and also damage tree roots (see Chap. 7 for comments on how to avoid compaction).

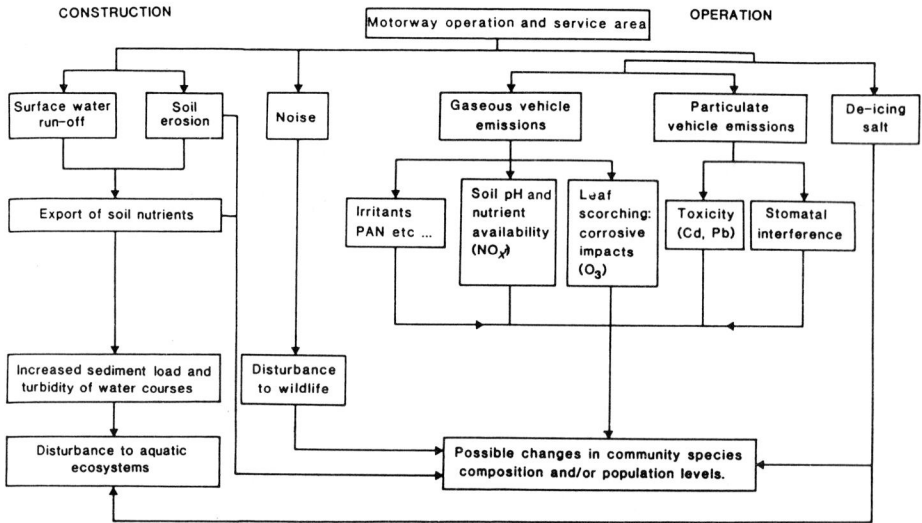

Fig. 5.1. Physical effects from motorway development (from Sheate and Taylor, 1990).

During construction, sites and sometimes associated base camps are set up for people involved in the project. The construction sites and base camps may be temporary but it is essential that their ecological effects be identified and addressed.

In preparation for the road, soil and rock may have to be excavated and transported elsewhere. Disposal of soil and rock from road excavations can have damaging ecological effects. It is not uncommon, in montane areas, for rock and debris to be simply dumped over the road edge. Loose scree and debris from such roads can cover and destroy more vegetation than the road itself.

In some instances base-rich rock material has been deposited on acid-rich soils and vice versa. The flora of acid-rich communities is quickly destroyed by base material.

Hydrological and geological effects are common, particularly where excavations into hill- or mountainsides have been made. That and the need to infill can result in many changes to the hydrology and geology of the area. Such changes have long been identified (Parizek, 1971) and remain the same today:

— cutting off tops of aquifers,
— development of extensive groundwater drains,
— damage to and pollution of catchments,
— changes in groundwater and surface water,
— sedimentation,
— siltation of channels causing flooding and erosion,
— obstruction to groundwater flow,
— changes in runoff and recharge characteristics.

In some cases roads are constructed across small stream channels and the channels ignored if at the time of construction the stream is dry or alternatively, culverts installed that are too small.

Road design, including grading, drainage and road alignments are all common to road construction. A so-called 'ecologic' (not ecological) approach to the 'principles and practices of grading, drainage and road alignment' has been described in a book of the same name by Untermann (1978). It is an introductory text intended largely for civil and construction engineers and landscape architects. Laying out roadways and determining when not to grade are treated in some detail. Despite the term 'ecologic' in the book title, ecology in the sense of interactions between roads and wildlife is hardly apparent as in most texts on road construction, engineering and landscape architecture are the main considerations in this one also.

Road cuttings are common in road construction and much has been written about the best profiles for soft-rock cuttings and hard-rock cuttings. Cuttings have a visual impact and their revegetation or offsetting impacts arising from erosion pose challenges. However, there are some interesting implications for geological conservation. Road construction can enhance geological and geomorphological resources whether for scientific use, education or for local interest. Construction of some roads presents opportunities for examination of classic geological sequences. A discussion document written by Larwood and Markham (1995) examines the best practical and technical solutions available for geological and geomorphological conservation on road cuttings.

In some parts of the world, such as the Himalayas, road construction (and deforestation) have been ranked among the most serious of all human impacts

(Haigh, 1994). In the Himalayas, road construction causes forest removal both directly and indirectly. Haigh (1982) estimated that clearance during road construction may have eliminated 1/240 of the trees in the Uttar Pradesh hills. Furthermore, the increased access provided by roads increased tree theft (Haigh, et al., 1995). Tree theft is a major cause of forest loss in parts of the Himalayas. Landslide activity also causes forest degradation.

Some roads, such as forest plantation roads (or logging roads) are temporary but nevertheless the resultant soil compaction, effects of ballast and other introduced materials, and erosion can have considerable short-term and long-term effects. Initial damage to forests adjacent to road construction can be significantly worse than that caused by subsequent secondary roads. Depending on the method of road construction, the direct loss of trees for the road can be significant. For example, a road 10 km long would require removal of many hectares of forest. Other trees are subsequently killed as a consequence of ground compaction, changes in levels of soil moisture, dumping of spoil, physical damage to roots and windthrows. The subsequent introduction of invasive weed species has been of considerable concern in many areas where roads have been constructed in forests.

Most research has centred on the effects of logging roads (and how to minimise these effects) in temperate climatic regions and until recently studies on the effects of logging roads in tropical forests were few. A study undertaken in 1993-1994 in the Central African Republic by Malcolm and Ray (2000) found that edge effects along roads continued to increase 19 years after construction, thus highlighting the likelihood of limited forest regeneration. The authors also found that the ecological effects of secondary logging roads (typically abandoned after logging) were as strong as those from primary logging roads. They suggest that damage to forest canopies needs to be reduced and present several methods for achieving this. Taking advantage of existing gaps for felling trees is one possibility.

Studies undertaken before, during and after construction of roads are few. One notable example is Interstate Highway 95 (I-95) in northern Maine, USA. From 1975 onwards, several people assessed the impact of this highway on the distribution, abundance and diversity of birds, rodents and other mammals (Burke and Sherburne, 1982; Sherburne, 1985). Loss of habitat had an impact on birds and mammals and an increase in abundance of edge species associated with construction of the road edge was noted. There appeared to be no significant impact on breeding bird populations in adjacent forests (based on comparisons between the preconstruction phase and the post-construction phase), but bird behaviour was affected by general disturbance and noise during construction.

Construction impacts need to be managed and management presupposes research. Such research was undertaken for example on the effects of temporary ballast roads on heathland vegetation in southern England. Rose and Webb (1994) carried out a series of trials and found that heathland plots disturbed early in the growing season regenerated more rapidly than those disturbed later in the year. Furthermore, although ten invasive plant species had established, none were in large enough numbers to alter the original vegetation.

Some roads are designed with high kerbs and large drains. These can prove fatal to amphibians trying to cross roads during periods of migration. A report in BBC Wildlife Magazine for January 1997 noted that high kerbs may prove insurmountable for some species and not uncommonly toads fall into gulley pots and drains and drown.

EROSION

Road construction accelerates landslide activity. Landslides are responsible for a very large amount of road-related sediment production. In the Himalaya, it has been suggested that the average sediment outfall per km of new hill road may be as great as 550 m^3 with some in excess of 1100 m^3 per km (Haigh, 1984). Road-bank erosion and landslides caused by road construction can be seen throughout many thousands of km in many parts of the world. In much of western and south-western USA, road-bank erosion is a conspicuous problem. Although road-bank erosion can damage road structures and surfaces as well as the adjacent environment, only in the last ten years have systematic studies and inventories of road-bank erosion been published.

During road construction severe problems may arise with erosion and sediment from earth-moving activities, leading to significant consequences for aquatic systems and wildlife. The construction phase can be particularly damaging but even after the road is completed, sediment may continue to enter waterways. Heavily used roads may contribute 130 times as much sediment compared to an abandoned road (Reid and Dunne, 1984). Sediment from paved roads may be as much as 1% that of gravel roads. In Scotland, construction of motorways during the mid-1990s led to serious pollution of rivers due to contamination with mineral soils and sediment. The contaminated rivers were part of a catchment containing designated salmon rivers. The offenders were therefore prosecuted and measures later adopted to avoid such impacts (McNeill, 1996).

Roads are an integral part of forestry activities. In the Pacific Northwest USA and elsewhere, forestry activities, in particular logging are known to accelerate debris flow and sediment production. Many reports (see for example Burns, 1972; King and Tennyson, 1984) have highlighted alteration of stream flow characteristics due to road construction and degradation of water quality and decrease in fish survival consequent to excessive sedimentation. Both salmon and trout are particularly at risk from debris and sediment caused by logging operations. The potential influence of forestry activities on water quality is growing rapidly and the most serious impacts occur during and after timber harvesting. Modern harvesting techniques and careful management of riparian habitats may help to moderate these impacts.

GENERAL DISTURBANCE EFFECTS

Roads and road traffic may disturb wildlife in a very general manner. Some birds in particular may avoid roads and be affected by a combination of effects. In a re-analysis of some previously published data, Van der Zande et al. (1980) looked at the general impact of roads on the population densities of four bird species in an open field habitat. For some species, such as lapwings (*Vanellus vanellus*) and godwits (*Limosa limosa*), there seems to be a general pattern whereby there are lower densities near roads over distances from 200 to 2000 m. The author concluded that long-distance disturbance effects must be considered in road environmental impact assessments.

The distances of bird nests to different types of edges needs to be considered when assessing bird nesting success in areas where there are roads in forested areas. In Pennsylvania, Yahner and Mahan (1997) examined the destruction of artificial ground nests in relation to logging-road width and the distance of nests from the logging roads. They compared predation rates on nests placed at induced edges created by narrow logging roads versus clear-cut edges. They concluded that predators were not using logging roads as wildlife corridors or concentrating foraging activities along these linear habitats.

Noise, Light and Wind

Noise and artificial lighting have been shown to affect some forms of wildlife. For example, noise and visibility of cars have been suggested as important disturbance factors for birds (Reijnen and Foppen, 1994; Reijnen et al., 1995). These authors analysed the population densities of 43 species of breeding birds at paired woodland sites close to and distant from busy roads. They found reductions in bird population densities (60% of the species showed reduced density). Species found to be most affected were buzzards, pheasants, woodcock, cuckoo, wood pigeons and some woodpeckers. For roads with traffic densities of about 10,000 cars a day, the effect on birds was apparent up to 1.5 km from the road. This distance increased to 2.8 km for roads with traffic densities reaching 60,000 cars per day.

The noise created by traffic depends on many factors, viz. nature of the road surface (sealed or gravel), traffic densities and types of traffic. It is likely that some traffic noises interfere with vocal communications between individual birds and could therefore interfere with courtship behaviour. An area worthy of research would be the effects of noise from different types of roads with different traffic densities.

Some wildlife are attracted to artificial lighting and the timing of behaviour of some birds could be altered. Although there is much research on the effects of noise and light on nature, the effects of road traffic noise and artificial lighting has not been well researched.

The effects of traffic wind gusts on leaf growth of various trees and shrubs living on the borders of motorways were studied in Switzerland by Fluckiger et al. (1978). The authors observed that plants grown near a motorway showed in general smaller leaves and necrotic leaf areas. About 30,000 vehicles passed the plants daily. It is suggested that plant disturbance caused by traffic wind gusts may be responsible for reduced leaf growth.

Dust and Sand

Apparently, few studies have been done on the chemical and physical effects of road dust on natural and seminatural biological communities. Some of these few studies relate to specific regions or biomes such as the tundra and taiga (see for example Forbes, 1995; Walker and Everett, 1987). The effects of aeolian sand and dust (from roads and quarries) on tundra seem to be due in part to changes in surface albedo resulting in changes in vegetation composition and vegetation cover. Forbes (1995) found dramatic changes in plant community composition and cover up to 200 m downwind from a typical sand quarry. One of the dominant plants in the mires of the Yamal region, Siberia, is *Rubus chamaemorus*. Its fruit is favoured by the aboriginal Nenets. Forbes, Siberia, found 62% fewer berries than in similar areas were recorder within 35 m of eastern edges of the road/rail corridors.

The influence of airborne road dust on the chemistry of Sphagnum mosses has attracted some attention. Physical effects may include cell destruction and blocked stomata. Chemical effects may arise from elements such as Al, Cr, Fe and Ni deposited in airborne road dust and affecting biota via soil enrichment (see for example Santelmann and Gorham, 1988).

A recent and comprehensive literature review by Farmer (1993) describes the effects of dust types on crops, grasslands, heathlands, trees, arctic bryophyte and lichen communities. Dust may affect photosynthesis, respiration, transpiration and facilitate affects of gaseous pollutants. Farmer concluded that epiphytic lichen and Sphagnum dominated communities were the most sensitive of the groups studied.

Most studies on effects of dust and sand on vegetation have been undertaken without quantifying patterns and extent of dust deposition. An application of SPOT satellite imagery could help to extend studies on effects of dust on vegetation. Keller and Lamprecht (1995) applied a simulation model for atmospheric diffusion and dry deposition of coarse particles to compute the dispersion and deposition of dust generated from the Dalton Highway in Arctic Alaska. The satellite-based method is potentially valid for quantising pollutants of this kind and indeed may be the only way to detect long-term effects.

Metals

Of all the physical and chemical effects of traffic and roads, the effects of metals and especially heavy metals have been researched far more than others (Table 5.2). Lead contamination of roadside ecosystems was extensively reviewed by Smith (1976) and later by Scanlon (1987, 1991).

106 Ecological Effects of Roads

In India, the dispersion of lead from vehicle exhausts was studied by Veerabhadra Swamy and Lokesh (1993). They reported that concentrations on roadsides decreased with distance and that lateral distribution seemed to be reduced by roadside trees. Penetration of lead into the soil was detected up to a depth of 15 cm.

Analysis of dust from sites along the M6 motorway in England has revealed very worrying levels of metals. The M6 carries huge volumes of traffic, in the order of 80,000 vehicles per day. A study (University of Wales) commenced in 1996 has shown that several metals exceed the threshold limits set in 1987 by an Interdepartmental Committee on the redevelopment of contaminated land. All the metals come from vehicles and include nickel (diesel fuel), zinc, cadmium, copper, chromium and lead.

Elements such as Pb, Ni, Cd, and Zn, which arise from petroleum products and car tyres, find their way into road residents and roadside biota. These elements are known to accumulate in roadside plants, animals and other organisms. Some road residents such as pigeons (which utilise bridges as nesting and roosting sites) have been studied as monitors of pollution from roads and traffic (Hamilton and Harrison, 1991). Other indicators used in high emission areas are the levels of heavy metals in tree bark and roadside mosses.

Research on traffic-related heavy metals and wildlife was particularly popular in the late 1970s and the 1980s. A few examples are given in Table 5.2. Then, and even today, by far most of the research centres on levels of metals in body tissue and not on the effects on populations and communities, and certainly not on long-term effects in ecosystems.

Table 5.2: Examples of literature on traffic-related heavy metals in roadside biota (see Reference at the end of this book for full bibliographic details).

Ash and Lee. 1980. Lead, cadmium, copper and iron in earthworms from roadside sites.
Atkins et al. 1982. The evolution of lead tolerance by Festuca rubra on a motorway verge.
Backhaus and Backhaus. 1987. Distribution of long range transported lead and cadmium in spruce stands affected by forest decline.
Beeby. 1985. The role of Helix aspersa as a major herbivore in the transfer of lead through a polluted ecosystem.
Chmiel and Harrison. 1981. Lead content of small mammals at a roadside site in relation to pathways of exposure.
Flanagan et al. 1980. Deposition of lead and zinc from traffic pollution on two roadside shrubs.
Foner. 1987. Traffic-related pollution of some edible crops in Israel.
Gish and Christensen. 1973. Cadmium, nickel, lead and zinc in earthworms from roadside soil.
Greszta. 1982. Accumulation of heavy metals by certain tree species.

Contd.

Table 5.2: (Contd.)

Grue et al. 1984. Lead concentrations and reproduction in highway-nesting barn owls.
Grue et al. 1986. Lead concentrations and reproductive success in European starlings *Sturnus vulgaris* nesting within highway roadside verges.
Hassel et al. 1980. Heavy metals in a stream ecosystem at sites near highways.
Isermann. 1977. Method to reduce contamination and uptake of lead by plants from car exhaust fumes.
Krause and Kaiser. 1977. Plant responses to heavy metals and sulphur dioxide.
Laaksovirta et al. 1976. Observations on the lead content of lichen and bark adjacent to a highway in southern Finland.
Lagerwerff and Specht. 1970. Contamination of roadside soil and vegetation with cadmium, nickel, lead and zinc.
McCreight and Schroeder. 1977. Cadmium, lead and nickel content of *Lycoperdon perlatum* Pers. in a roadside environment.
Mukherjee et al. 1998. Blood lead levels in traffic personnel in Calcutta.
Muskett and Jones. 1980. Dispersal of lead, cadmium and nickel from motor vehicles and effects on roadside invertebrate macrofauna.
O'Neill et al. 1983. Lead contamination near Kansas (USA) Highways: implications for wildlife enhancement programmes.
Quarles et al. 1974. Lead in small mammals, plants and soil at varying distances from a highway.
Rodriguez-Flores and Rodriguez-Castellon. 1982. Lead and cadmium levels in soil and plants near highways and their correlation with traffic density.
Scanlon. 1987. Heavy metals in small mammals in roadside environments: implications for food chains.
Schafer et al. 1998. Uptake of traffic-related heavy metals and platinum group elements (PGE) by plants.
Smith. 1976. Lead contamination of the roadside ecosystem.
Udevitz et al. 1980. Lead contamination in insects and birds near an interstate highway, Kansas (USA).
Veerabhadra Swamy and Lokesh. 1993. Lead in roadside soils in India.
Wade et al. 1980. Roadside gradients of lead and zinc concentrations in surface-dwelling invertebrates.
Ward et al. 1974. Effect of lead from motor-vehicle exhausts on trees along a major thoroughfare in Palmerston North, New Zealand.
Ward et al. 1975. Lead in soil and vegetation along a New Zealand state highway with low traffic volume.
Ward et al. 1979. Seasonal variation in lead content of soils and pasture species adjacent to a New Zealand highway carrying medium density traffic.
Wheeler and Rolfe. 1979. Relationship between daily traffic volume and distribution of lead in roadside soil and vegetation.

In one study in Germany, the highest levels of heavy metal accumulation were found to occur in mosses on the ground (Lotschert and Kohm, 1978). This seems a well-researched area with respect to levels of and rate of accumulation in relation to traffic volumes (see for example Motto and Daines, 1970). The effects of accumulation of heavy metals in biota have prompted less attention.

Some authors have noted that some organisms seem not to be harmed despite accumulation of metal elements. In fact, some plants have evidenced enhanced root growth as a result of soil contamination from roadside dust carrying trace metals (Wong et al., 1984). Among invertebrates, the effects may or may not be detrimental. For example, Przbylski (1979) found that whereas combustion gases may cause reduction in species richness in arthropods, some groups flourish in such an environment. In another earlier study, Muskett and Jones (1980) found no obvious decline in either numbers of invertebrates caught in pitfall traps or in species diversity (alpha diversity index) with increase in metal pollution load.

Some research has drawn attention to the security of research on the effects of metals consumed by animals feeding on biota near roads. Concern about bioaccumulation of metals from one trophic level to another has been expressed (see for example Scanlon, 1987). There have also been warnings about possible effects on humans who may consume roadside plants and fruits (Rodriguez-Flores and Rodriguez-Castellon, 1982). Some studies have shown that the levels in animal predators are too low to be toxic (Wade et al., 1980). One study has suggested that contamination of roadside habitats by lead does not pose a serious threat to highway nesting birds that are aerial feeders, such as swallows (Grue et al., 1984).

Gases

In many countries, transportation (traffic on roads and ships) is a key source of nitrogen deposition. In 15 member states of Europe, transportation accounts for almost two-thirds of all the nitrogen oxides emitted (Agren, 1996). In Germany alone, half of the annual 3.1 million tons of NO_x emitted is from traffic (Bach, 1985). It is known that nitrogen compounds and ammonia are a major cause of eutrophication and it is now generally considered that NO_x and HC are the main ingredients in the photochemical oxidant formation which contributes to forest dieback.

Some species-specific studies have been done. For example, Spencer and Port (1988) found that *Lolium perenne* grows more vigorously in soil taken from near roads and considered NO_x and de-icing salt possible causes. Sarkar et al. (1986) contrarily recorded stunted growth in plants near highways in Calcutta. Kammerbauer et al. (1986) drew attention to the apparent lack of experimental work on the effects of exhaust emissions on forest trees (noting the earlier controversy over the reasons for dieback in forest trees near roads). They presented evidence that injuries of Norway spruce are caused by exhaust emissions of CO and NO_x). They also demonstrated drastic reductions in these effects when catalytic converters were used.

Angold (1992) in his detailed study of the ecological effects of road pollutants on heathland communities in southern England, found that nitrogen augmented plant growth (effects were measured up to 200 m into the heathland) and subsequently the species composition changed. In Poland, a study by Przbylski (1979) recorded changes in abundance of groups of arthropods living among cultivated plants (winter wheat and apple orchards). Based on these results and previous research on the effects of industrial pollutants on arthropods, he concluded that:

- Traffic-emitted gases do reduce the number of species of arthropods in orchards and crops.
- Arthropods of the family Aphididae and members of Heteroptera flourish.
- Some Lepidoptera and members of the family Thripidae families are affected by reductions in numbers.
- The selective effect of combustion gases is associated with feeding mode, anatomy and life history of the insects.

Runoff (Effects on Aquatic Systems and Biota)

Effects of roads and runoff (containing heavy metals, nutrients and other contaminants) on aquatic biota and aquatic ecosystems have attracted much attention. The effects from both road construction and operational roads have been investigated with respect to impacts on nature. In general, the major effects of roads on streams are influx of debris and sediment that severely disturbs the stream flow or removal/rearrangement of material in the streambed (Jones et al., 2000).

Serious river pollution has occurred in some instances during road construction (McNeill, 1996). Extence (1978) has indicated that discharges resulting from road construction can be serious enough to warrant implementation of control measures.

The effects of pollutants (in water runoff from roads and particularly those in urban areas) on aquatic biota and ecosystems include both immediate and long-term effects (the former seems well researched but not the latter). Water run-off from roads alters the hydrology, may increase sediment load, increase nutrients and also result in accumulation of many kinds of pollutants.

Increases in sediment load and changes in stream flows resulting from logging activities has long caused concern. In 1972, Burns noted sustained logging prolonged adverse conditions in streams for fish. Eaglin and Hubert (1993) found that increased sediments in streams resulting from logging operations had a deleterious effect on fish populations.

The design of retention or detention ponds receiving highway run-off has for many years been based on hydrology and hydraulic considerations only. The effectiveness of retention ponds for removal of contaminants has only recently been researched.

The role of wetlands as 'sinks' for metals and macronutrients from roads has also been much researched (see for example Yousef et al., 1996). Conflict occurs here; on the one hand avoidance of pollution of wetlands is attempted and, on the other, wetlands are used as pollution sinks.

A particularly detailed study on the effects of motorway runoff on freshwater ecosystems was undertaken by Maltby et al. (1995). They reported changes in species diversity and composition of macroinvertebrate assemblages but found no changes is either diversity or abundance of epilithic algae.

According to Dickson (1986), the most damaging agent in aquatic habitats is siltation and increase in nutrient loads rather than the introduction of chemicals.

Runoff from roads in urban areas is of course exacerbated by the area of impervious surfaces. Recognition has dawned that the extent of impervious surfaces and compacted ground is a significant contributory factor to flooding. Since the impervious surfaces of urban and industrial areas, roads and compacted agriculture soils greatly determine the rates and volumes of water flow, it is not surprising that urban waterways are often no more than large drains designed to move water away as quickly and safely as possible. The trouble is that many of these drains are no longer able to cope with the volume of water and, ironically, there is a trend these days to restore natural waterways and thereby regain the natural services of nature (Fig. 5.2). More generally speaking, however, there is now sufficient evidence that river engineering can lead to physical destruction (see for example Jeremy Purseglove's book *Taming the Flood* published in 1989).

De-icer Agents

Salt (especially rock-salt) mixed with sand is the most commonly used agent for de-icing roads. Thousands of tons of salt (millions of tons in some parts of the world) are applied each winter to many roads in northern Europe and northern United States, Canada and Alaska. The salt leaches through the soil and salt spray deposits on plant surfaces. Salt in roadside soils may also increase the available nitrogen and some minerals (Spencer et al., 1988; Townsend, 1984). The salinity of motorway soils and subsequent fate of the chemicals and effects on shrub species have been extensively reported in a series of papers by Thompson et al. (1986).

The skin of amphibians is thin and permeable and makes them vulnerable to irritants such as salt. Consequently it is not surprising that toads crossing roads which have been salted may suffer discoloration of their skin and some have been found dead. In the UK, road salting takes place about the same time that toads migrate.

The extent of harm to plant species depends on the sensitivity of the plant species and levels of applications. By the end of winter, many roadside verges are devoid of any vegetation. Repeated applications of salt have led to changes in plant species composition in some locations (Davison, 1971) and acted as an agent in facilitating dispersal of halophytic species (Scott and Davison, 1985). Increased levels of salt uptake in some tree species may also lead to increased sensitivity to infection by fungus.

Much concern has been expressed about the effects of de-icing agents on the ecology of roadside plants. Surveys and experimental work on the effects of de-icing agents have been undertaken since at least the late 1960s and early

Fig. 5.2. Contrasting roadside stream management in Christchurch, New Zealand (photo, Ian Spellerberg).

1970s (see for example Westing, 1969; Davison, 1971). The effects on roadside verge vegetation may affect vegetation up to several metres from the road (Westing, 1969). Water runoff from the road could result in more distant effects in watercourses but little research has been done on this aspect (Good and Grenier, 1994). However, Forman and Deblinger (2000) report that salt from roads can reach streams in sufficient quantities to affect downstream aquatic ecosystems.

Perhaps the most widespread effect of salt as a de-icing agent is the increase in salt-loving (halophytes) plants and consequent change in species composition in favour of halophytes. In some cases, halophytes constitute the dominant vegetation.

Many coastal species are halophytic or at least salt-tolerant and coastal plant species have long been observed on inland roads in Europe and the UK. The most successful species is the grass *Puccinellia distans* and indeed, in Britain, it is the most widespread plant on roadsides (Scott and Davison, 1985). Other species reported by Scott and Davison to invade roadsides include *Aster tripolium, Cochlearia officinalis, Halimone portulacoides, Plantago maritima* and *Suaeda maritima*.

Calcium magnesium acetate (CMA) is used as an alternative to NaCl for de-icing highways. Reports in the USA from the 1980s suggest that CMA may have beneficial effects on trace metal mobility (Amrhein and Strong, 1990).

Herbicides and Pesticides

In some locations, management of roadside verges involves the use of herbicides and/or pesticides. Non-target species are easily affected. There seem to be few policies or controls over the use of chemicals on roadside verges.

Litter

Thousands of tons of litter are deposited on the edges of roads every year. The discarded litter can trap small animals such as shrews, mice and lizards (Fig. 5.3). Discarded cigarettes, bottles and glass are the cause of fires in roadside vegetation. In parts of southern England, litter on roadside edges has been deemed the cause of fires that have destroyed vegetation in nature reserves (Fig. 5.4).

The costs of clearing roadside litter are not negligible. Indeed, a review of the state maintenance of litter programmes in the USA by Andres and Andres (1995) shows these costs to be quite high. In 1993, the cost in California alone was $28 million and for 42 States the total was $131.6 million (Box 5.1).

Stock Truck Effluents

Spillages from road vehicles is one source of pollution. There are anecdotal reports of deliberate spillages from tankers carrying liquid waste when it is

Physical and Chemical Effects of Roads and Traffic 113

Fig. 5.3. Examples of dead small mammals found trapped in roadside litter in Europe (photo, Ian Spellerberg).

114 *Ecological Effects of Roads*

Fig. 5.4. Photographs taken in Europe before and after a fire destroyed the vegetation on a nature reserve adjacent to a road in southern England (photo, Ian Spellerberg).

Box 5.1:

Costs (in 1991 or 1993) of clearing the roadside litter and percentage of maintenance budget for some states of the USA (from Andres and Andres 1995).

State	1993 Costs in Millions	% of Maintenance Budget
California*	$28.0	5.6%
Illinois	7.5	4.0%
Texas	5.5	1.5%
Florida	5.5	2.6%
Washington	5.4	5.0%
Kentucky	5.6	4.3%
Pennsylvania*	5.0	4.7%
Virginia	4.8	1.0%
Maryland	4.75	4.5%
New Jersey	4.7	8.8%
Oklahoma	3.6	5.0%
Ohio	3.3	3.1%
Missouri	3.0	1.0%
West Virginia	3.0	1.0%
New York	2.8	1.5%
Nevada	2.75	6.8%
Colorado	2.5	5.0%
Connecticut	2.5	2.0%
Michigan	2.5	1.5%
Wisconsin	2.3	2.0%
South Carolina	2.1	1.7%
Minnesota	2.0	1.5%

*Data for 1991

raining. Purportedly, the tap on the tank is left slightly open and the liquid waste deliberately dumped on roads.

Effluents spilled from vehicles carrying cattle and other livestock have been recognised as a serious issue. Stock effluent spillage may be a hazard and result in accidents. The spillages are also perceived as offensive by some road users. Some vehicles are fitted with effluent storage tanks but very often the capacity of the tanks is too small. Thull (2000) researched the political, administrative, policy and economics of the issue. He suggests that a strong industry code of practice coupled with a strategic network of dump sites would enable the industry to address the issue successfully.

QUANTIFYING THE EFFECTS

The ecological effects can be quantified in detail at certain levels. For example, the area of habitat lost to road development can be computed. The accumulation of heavy metals in biota can be quantified (quantisised). In general, a list of criteria can be used as a basis for the quantification. For example, Sorokovikova (1990) promoted the modification of an interesting approach to the use of several criteria for assessing impacts of transportation on the environment which had previously been discussed by Filina (cited in Sorokovikova). This method uses the following:

— length of road
— area of road
— traffic density (vehicles per day)
— levels of pollution t/km^{-1}
— growth in traffic (average annual growth in number of vehicles).

Also proposed was an indicator of 'disconnectedness' (separation of adjacent natural areas) caused by roads. However, the author does not propose a method for calculation of this component due to the lack of spatial analysis data (what area of habitat is too small for certain biota?).

On the basis of these criteria, it is possible to prepare maps showing the relative intensity of impacts (Fig. 5.5). This kind of approach is worthy of much more research and development because of the implications for planning, design and management of the ecological effects of roads.

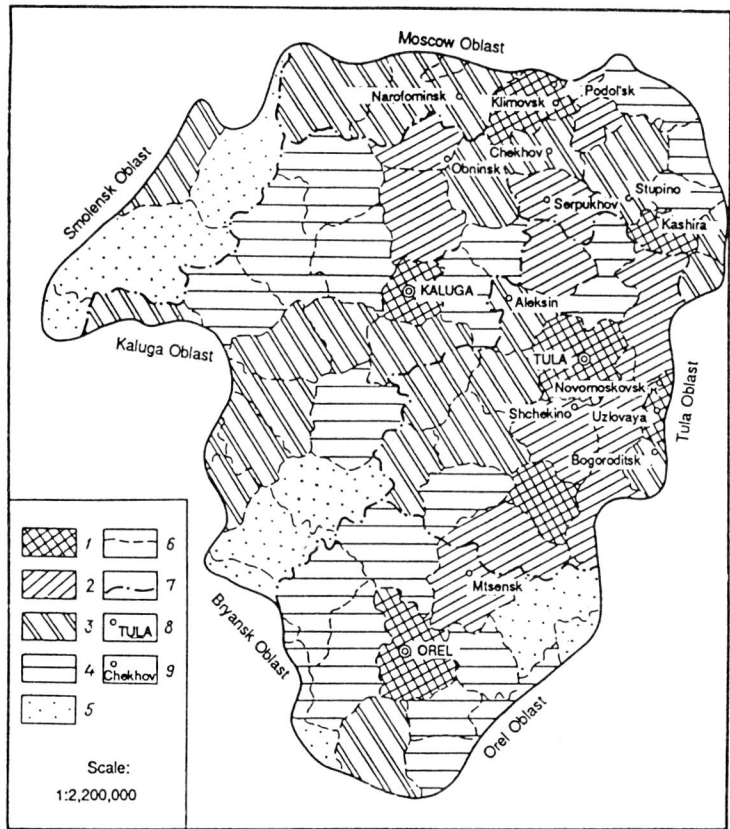

Fig. 5.5. Summary estimates of transport impacts on the natural environment of the Upper Oka basin (adjacent to and including the southern part of Moscow Oblast). The intensity of impacts is given from 1-5 with 1 being the highest. The symbols from 6-9 are boundaries. From Sorokovikova, N.V. 1990. Reprinted with permission from Soviet Geography, 31(2) 116–125. Copyright V.H. Winston and Son, Inc. 360 South Ocean Boulevard, Palm Beach, Fl. 33480. All rights reserved.

6

Road Kills: Animal Mortality on Roads

INTRODUCTION

Vehicles on roads (and trains on railway lines) are the cause of animal mortalities. Seeing a dead animal on or beside a road (or railway line) is not an uncommon sight in any country of the world. Occasionally these are domesticated animals such as cats, dogs or sheep. Most animal mortalities on roads (and railway lines, especially in rural areas and forested areas) are wild animals, however: birds, mammals, reptiles and amphibians (Fig. 6.1). While a dead possum on a road in New Zealand evokes little concern, a dead elephant beside a road or railway line in India (Fig. 6.1) or a dead koala on a roadside in Australia is likely to prompt considerable distress. Interestingly, as mentioned earlier, the first prize for a world press photographic award (nature and environment category) was awarded in 2001 for a picture of a dead lizard's tail lying on a road in Australia.

Mortality of animals on roads has an ecological dimension, animal behaviour dimension, emotive dimension and a social dimension. Road kills have been a cause for concern for many years and featured in magazine articles, cartoons, serious research, museum displays and school projects. Collisions between vehicles and animals on roads can cause considerable damage to the vehicle and engender distress, injury or even death of the people in the vehicle.

In the early 1900s, when there were only 750,000 miles of improved roads in the USA, around 38,000 people were killed in road accidents. Nor did the mortality of animals on roads go unnoticed. Perhaps the first and now a classic book about wildlife killed by traffic was that written by James Simmons, *Feathers and Fur on the Turnpike* which was published in 1938. Here, perhaps for the first time, wildlife mortality on roads was detailed and presented in a systematic and objective fashion. James Simmons' concern about the extent of animal mortalities prompted publication of a book that would reach as wide an audience as possible. The introduction to his book clearly expresses his concern (Box 6.1). Had a systematic monitoring programme been established at that time, perhaps today we would have an even more valuable record. Nonetheless the literature on animal road mortality is considerable (Table 6.1).

The mortality of animals on roads is discussed in this chapter. Logically, the first question is: Why are there animals on roadways? Then follow the questions: What is the extent and magnitude of animal road mortalities? Do road kills affect wildlife population dynamics? What can be done to reduce the level of animal road mortalities.

WHY ARE THERE ANIMALS ON ROADS?

Apart from domestic animals, which are occasionally and deliberately taken along roads or across roads, why are there animals on roads? Some roads are established across historical migratory routes of some mammals such as elephants (Fig. 6.2) and, in many cases, roads are included in the home ranges of some species of mammals. Large mammals, such as deer, cross roads (and railway lines) because the road is within the area they inhabit. Similarly, small mammals and marsupials (typically hedgehogs and species of possums) and other small animals such as amphibians, snakes and lizards cross roads as part of their normal dispersal and day-to-day movements. Scavengers frequent roads because of the available food sources.

During winters in temperate climatic regions, snow-covered ground may encourage some large mammals to use roads. However, some mammals such as reindeer are sometimes killed during winter months when they use roads as conduits for easier travel. Birds sometimes seek grit on roads during periods of snow cover. Water-birds are sometimes found on roads during periods of snow cover and it is possible that a road in snow-covered country, in a bird's-eye view, may look like a river.

At night, the glare of headlights confuses some wildlife and this may lead to increased road mortalities. Well-lit roads may extend the feeding time of birds on some roads. Some birds roost on roads at night because of the heat retained by the road surface. During the day, some birds dust-bath on road edges and the gravel and debris on road edges provides a source of grit for other species. Reptiles, especially lizards, are attracted by the warm surface of the road (Chap. 3, Fig. 3.1a).

Some animals are attracted to roads because of food such as insects flying about them, insects basking on the warm surfaces and dead animals lying there. Stoats and related species feeding on road carrion may themselves be killed. Hawks and owls feeding on or near roads are not uncommon. Kestrels feeding on road carrion are a common sight along many motorways in Europe. Grain and fruit spilt on roads attract some birds and small mammals. The plants of road verges may also attract some birds to feed on fruits; road kills may result. For example, Cedar waxwings (*Bombycilla cedrorum*) are killed on roads while feeding on fruits of roadside shrubs (Dowler and Swanson, 1982). In some locations the salt, used as a de-icing agent, collects in pools and attracts some animals such as moose.

The conditions and vegetation on road edges attract some animals for feeding and they become potential casualties. Both domestic and wild animals

(a) Pukeko in New Zealand

(b) Carcass of an elephant killed May 2000 in a train accident in Rajaji National Park, India (photo, Ajay Pal Singh).

(c) Grass snake on a read in Germany (photo Ian Spellerberg)

(d) Possum on a New Zealand road (photo Ian Spellerberg)

Fig. 6.1. Dead animals on roads and near railway lines.

> **Box 6.1:**
>
> Extracts from Simmons (1938) *Feathers and Fur on the Turnpike*.
> Extracts from the Introduction
>
> **Introduction to Part I**
>
> Most of us who drive automobiles are aware of the fact that a great number of animals lose their lives on the highway. From personal observation, and from short articles that have appeared in some of the scientific magazines, we can draw conclusions as to the number of casualties per mile.
>
> Not many of us have gone so far as to determine whether the killing is confined to rabbits, skunks and an occasional game bird, or whether it includes a cross-section of the entire wildlife population.
>
> Scientists will eventually give us the entire facts. They may or may not suggest some remedial measures. The chances are that losses in human life will continue to overshadow in importance these losses in wild animal life—to the extent that a safety campaign on behalf of the latter would be laughed out of court.
>
> I assume, however, that the general interest in wildlife conservation is sufficient at the present time to justify a report in book form on a ten-year study of animal victims of the automobile.
>
> **Introduction to Part II**
>
> There was never any question about the title of this book after I began recording observations on wildlife killed by motor vehicles on the road. But several chapters were completed before it occurred to me that the same title might be applied to a subject of greater significance. I refer to that modern movement or trend, national in scope, under which wildlife may eventually be restored to some measure of its former abundance in the land, and repossessed with some of its original dignity and glory.
>
> Wildlife conservation is marching down the highway to a long delayed recognition and honour. This is not because a large number of birds and mammals, dying on the turnpike, have focused attention on the subject. It is because we have been as casual in regard to the welfare of wildlife for two hundred years as we now are in regard to the escape or destruction of any small bird that happened to get in the path of our modern transportation.

such as deer and marsupials are common casualties because of their habit of feeding on road edges.

The edges of roads with strips of grass, embankments covered in shrubs and ditches with water provide suitable habitats for some species. Some reptile species in Europe have colonised south-facing road embankments and some species of amphibians make use of damp or water-filled roadside ditches.

For some species we might expect that the noise of traffic would deter their approaching a road. Furthermore, for some species it seems possible that

Table 6.1: Examples of the literature on wildlife road mortality

Web pages

Critter crossings
http://www.fhwa.dot.gov////////environment/wildlifecrossings/main.htm
http://www.tfhrc.gov////////pubrds/marapr00/critters.htm
Road kills of kangaroos. htp://www.act.gov.au/environ/KAC3/Kac3-APE.html
Highway Safety Information System (HSIS)
http://www.bts.gov/NTL/DOCS/hsis/94-156.htm

General literature and reports (see references at the end of the book for bibliographical details).

Case, R.M. 1978.
Cook, K.E. and Daggett, P-M. 1995.
Dickerson, L. 1939.
Evink, G.L. et al. 1996.
Forman, R. and Alexander, L. 1998.
Forman, R. and Deblinger, R. 2000.
Jonkers, D.A. and de Vries, G.W. 1977.
Lalo, J. 1987.
Manderville, V. and King, D. 1995.
McLure, H. 1951.
Simmons, J. 1938.
Smith, D.J. 1995.
Trombulak, S. and Frissel, C. 2000.

Invertebrates

Naumov in Nankinov, D. and Todorov, N. 1983.
Riffell, S. 1999.
Samways, M. et al. 1997.

Amphibians and reptiles

Boarman, W.I. and Sazaki, M. 1996.
Jackson, S. 1996.
Langton, T.E.S., 1989.

Birds

Brown, R. et al., 1986.
Finnis, R. 1960.
Hodson, N. and Snow, D. 1965.
Ducks: Sargeant, A. 1981.
Magpies: Brown, R. et al., 1986.
Owls: Kerlinger, P. and Lein, M. 1988.
Owls: Massemin, S. et al., 1998.
Pelicans: Owens, L.K. and James, R.W. 1991.
Partridge: Illner, H. 1992.
Swamphens: Washington, C. 2000

Contd.

Table 6.1: (Contd.)

Mammals

Bellis, E. and Graves, H. 1971.
Brockie, R. 1960, 1999.
Coulson, G. 1982.
Jefferies, D. 1975.
Madsen, A.B., 1996 (otters).
Mallick, S. et al., 1998.
Manen, F.T. et al. 1995.
Morris, P. and Morris, M. 1988.
Smith, D.J. 1995.
Williamson, L. 1980.

traffic noise could impact on vocal communications. Illner (1992) has suggested that energy in the form of noise, vibration or visual stimuli produced by moving vehicles could be a major factor contributing to the reduced population density of grey partridge (*Perdix perdix*) near roads. He further suggests that loud and permanent traffic noise could reduce the efficiency of contact and alarm calls and the capacity for detecting predators.

The risk of road kills increases at various times during a species annual cycle of activity. For example, in the Everglades of Florida, the greatest period of activity of some snake species coincides with peak periods of visitors to national parks. This results in a high mortality with 73% of all snakes on roads being killed or injured. For some species the greatest incidence occurs during migratory movements and during breeding seasons. Frogs and toads attempting to cross roads at the commencement of the breeding season are commonly killed in higher numbers.

WHAT CAUSES THE CASUALTIES?

In other words, what characteristics of roads and traffic contribute most to road kills?

That many animals are killed on roads each year is a fact. What contributes most to those deaths is not well known. Perhaps those roads with more traffic have higher road kills. Perhaps road kills are related to traffic density. Analysing the causes of road kills could provide information for developing ways and means for reducing the level of road kills. In one study of road kills in Nebraska it was found that except for badgers, road kills did not correlate significantly with average daily traffic but did correlate significantly with average vehicular speed (Case, 1978). Other studies have confirmed that traffic speed is the primary factor in animal kills on roads.

In a detailed study in Ottawa, Canada on the effect of road traffic on amphibian density, Fahrig et al. (1995) found that the combined number of dead and live frogs and toads per km decreased with increase in traffic density while the proportion of dead frogs and toads increased.

Some sections of roads are prone to a higher incidence of road kills. For example, topographical features may obscure a driver's vision of animals on roads. Similarly, such features may also interfere with or limit an animal's ability to detect oncoming traffic.

In the USA, Forman and Hersperger (1996) researched road ecology in different landscapes. Clare Washington (2000) in New Zealand researched the swamp hen or pukeko (*Porphyrio porphyrio melanotus*) commonly seen on roadsides and vulnerable to fatality. Many factors contribute to the mortality of wildlife on roads. These include road network density, road width, location of the road, traffic density and speed, nature of the surrounding habitats, driver behaviour and road conditions, and animal behaviour (seasonal, reproductive) and physiological condition.

WHAT IS KILLED? AND WHAT IS THE EXTENT OF MORTALITY?

Multiple surveys of mortalities of vertebrate animals on roads show that a wide range of species are killed on them every day (see Table 6.1). The list includes amphibians (especially frogs and toads), reptiles (in particular snakes), birds, monotremes, marsupials and mammals (from mice through armadillos to moose).

Mortality levels of reptiles and amphibians have been well documented for North America and Europe (Dodd et al. 1989; Langton, 1989). Many invertebrates are killed by traffic but no inventory on the extent of mortality is available.

Estimates of the extent and magnitude of animal road deaths have been attempted. Some estimates are based on records of actual road kills along certain lengths of roads for particular periods of time. Figures of around a million animals per day have been suggested (Lalo, 1987) and in Australia the number of frogs and reptiles killed annually on roads has been estimated at around five million. A wildlife information and rescue service has estimated that up to 12 million native animals are killed on Australian roads each year.

In the Netherlands, Jonkers and De Vries (1977) estimated the yearly average number of animal traffic casualties at 653,000 birds and 159,000 mammals. An astonishing annual mortality of seven million birds has been estimated for Bulgaria (Nankinov and Todorov, 1983). Barn owls (*Tyto alba*) seem particularly at risk from traffic and in England it has been reported that over half of all reported barn owl casualties are road deaths. Estimates put the number of birds killed on roads in the UK every year at 10 million. In the Netherlands, it has been said that 15% (400-500 individuals) of the total badger population is killed every year by traffic (van Bohemen, 1998).

Deer are particularly at risk. Williamson (1980) wrote that 200,000 deer were killed on roads in the USA in 1980. White-tailed deer are frequently killed on roads in some states of the USA and estimates suggest 37,000 annually. Romin and Bissonette's report (1996) is another more recent and

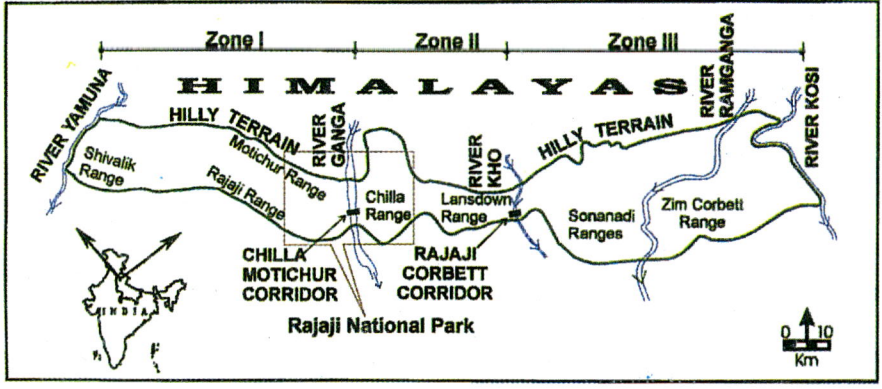

Fig. 6.2. (a) Rajaji-Corbett Elephant Conservation Unit.

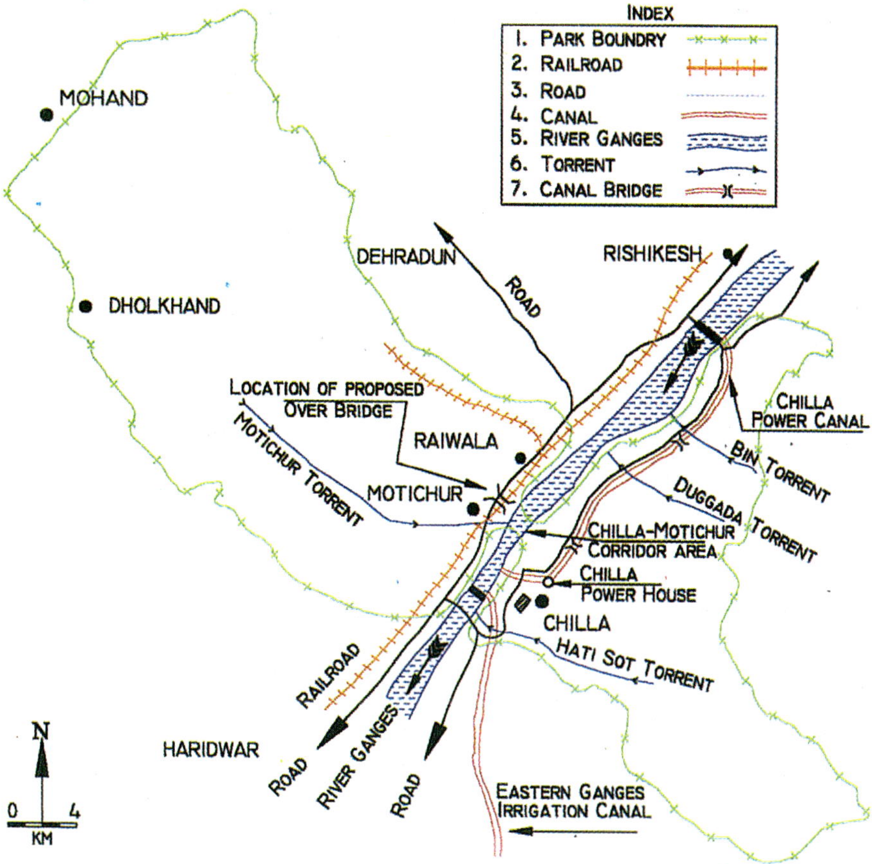

Fig. 6.2. (b) Alignment of Rail road, Highways (Road), Canals and Chilla-Motichur Corridor in Rajaji National Park.

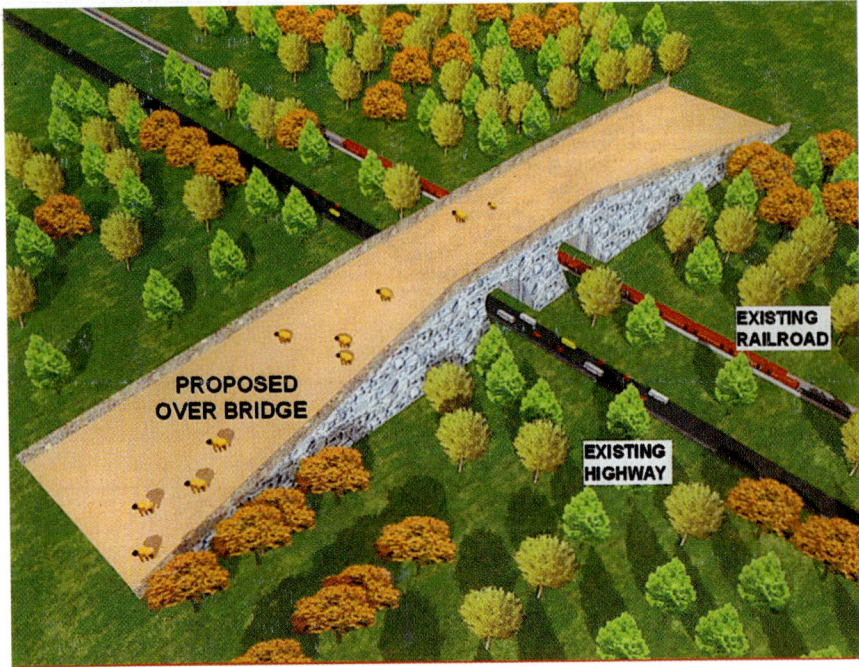

(c) Computer generated perspective view of proposed Eco-friendly overbridge on rail line and highway..

Fig. 6.2. Designing highways, railways and canals in protected areas to reduce human-elephant conflicts (from Ajay Pal Singh, 2000). A total of 16 elephants have been killed in the last 14 years on the Haridwar-Dehra Dun railway in Rajaji National Park, Uttaranchal, India.

Table 6.2: State-wise changes in deer mortality on USA highways from 1982 to 1991, (from Romin and Bissonette, 1996)

State	No. deer killed (year)				Actual Count made?	Per cent[a] change
	Lowest year		Highest year			
Alabama	no data					
Alaska[b]	51-77/year				no	
Arizona	no response					
Arkansas	3,603	(1990)	4,200	(1989)	no	(14)
California	15,000/year				no	
Colorado	5,202	(1983)	7,296	(1991)	no	40
Connecticut	1,429	(1982)	2,423	(1986)	yes	70
Delaware	103	(1982)	268	(1991)	yes	160
Florida	no data					
Georgia	50,000/year				no	
Hawaii	no response					
Idaho	no data					

Table 6.2: (Contd.)

State	No. deer killed (year)				Actual Count made?	Per cent[a] change
	Lowest year		Highest year			
Illinois	2,797	(1982)	15,560	(1991)	yes	456
Indiana	2,858	(1982)	12,671	(1991)	yes	343
Iowa	4,805	(1982)	9,248	(1988)	yes	92
Kansas	2,492	(1982)	3,536	(1991)	yes	42
Kentucky	1,490	(1982)	4,677	(1990)	yes	214
Louisiana	1,500/year				no	
Maine	2,000	(1980's)	3,500	(1990's)	no	75
Maryland	no response					
Massachusetts	no data					
Michigan	18,045	(1982)	44,374	(1991)	no	146
Minnesota	11,471	(1982)	16,280	(1991)	yes	42
Mississippi	no data					
Missouri	4,779	(1982)	9,519	(1987)	yes	99
Montana	no data					
Nebraska	1,261	(1982)	3,341	(1991)	yes	165
Nevada	no response					
New Hampshire	455	(1982)	1,000	(1990)	yes	120
New Jersey[c]	455	(1982)	10,494	(1986)	yes	2,206
New Mexico	no data					
New York	7,269	(1984)	10,978	(1991)	yes	51
North Carolina	5,000-8,000/year				no	
North Dakota	2,500	(1980's)	3,000	(1990's)	no	20
Ohio	8,587	(1982)	20,215	(1991)	yes	135
Oklahoma[d]	450	(1985)	495	(1983)	yes	(9)
Oregon	no data					
Pennsylvania	24,648	(1983)	43,002	(1990)	yes	74
Rhode Island	no response					
South Carolina	840	(1982)	3,689	(1991)	yes	339
South Dakota	2,166	(1982)	3,363	(1991)	yes	55
Tennessee	no response					
Texas	(19,000/year)				no	
Utah	1,826	(1980-81)	5,502	(1988-89)	yes	201
Vermont	1,105	(1982-83)	1,514	(1990-91)	yes	37
Virginia	1,446	(1982)	3,427	(1990)	yes	137
Washington	no response					
West Virginia	3,844	(1985)	9,515	(1991)	yes	148
Wisconsin	28,878	(1982-83)	76,626	(1989-90)	yes	165
Wyoming	987	(1988)	1,756	(1982)	yes	(44)

[a] Values in parentheses are decreases.
[b] Highest yearly estimate was used to calculate total deer kill in 1991.
[c] Actual counts through 1988-1989, estimated 10,000 deer/year from 1989-1992.
[d] Record keeping discontinued after 1985 due to unreliable efforts.

comprehensive account of deer mortality on highways (Table 6.2). Between 1982 and 1991 level of road kills had increased in most localities (many increases were significant).

What level of confidence can we have in these actual numbers of road deaths and estimates of road deaths. For some groups of animals, the removal of road kills is reflected in the statistics. It is known (although not to what extent) that reptile road deaths may be removed by reptile ecologists as part of biogeographical studies, thereby diminishing survey counts. Mammals are sometimes removed by hunters or collectors. Some animals may be fatally injured but die away from the road.

It is likely that many of these estimates are generous, even in intensive studies. Road deaths per se are of concern to many people. But do these deaths have any impact on the populations?

PUTTING THE CASUALTIES INTO PERSPECTIVE

Estimates of road kills appear alarming and indeed such loss of wildlife on the scale mentioned above is not pleasing. Nevertheless important questions to ask are what proportion of animals are killed and are there implications for the populations or the species (putting the numbers into perspective is not to deny the loss of wildlife).

For some species, road deaths do have serious implications for population levels. Examples in general are rare species and in particular those with fragmented distributions. For many other species, the numbers killed are probably not significant in terms of the total population. Bennett (1991) has suggested that populations of larger mammals are more affected than populations of small mammals.

Amphibians are particularly vulnerable to road kills when roads obstruct their traditional migration routes. In Wales (UK), movement of toads across roads and resultant mortalities have caused concern. Estimates of toad deaths have been put at 4% (compared to an annual mortality rate of 50-60%-Gittins, 1983).

Small mammals are very vulnerable to road kills but the effect on populations may be small. In Texas for example, less than 1% of rodents living on roadsides are killed per annum (Schmidly and Wilkins, 1977).

Ducks crossing roads result in road kills but again the effect on populations may be insignificant. For example, the road mortality of nesting ducks in North Dakota has been reported as less than 0.2% of the local annual breeding population (Seargent, 1981).

Studies on the implications of road casualties for animal populations, let alone species, are few. However, road kills may have some implications for the conservation of some species, especially when the population is small or scattered and local. Fahrig et al. (1995) after a detailed study of the effects of traffic on amphibian density concluded that road kills do have a significant negative effect on the local density of these animals. They add that traffic volumes could be contributing to worldwide declines in amphibian populations.

For endangered species of crocodiles (Kushlan, 1988), the garter snake (Dalrymple and Reichenbach, 1984) some species of deer, black bears and the bald eagle in Florida, road kills are thought to be an important contributing factor to mortality. Deaths of black bear (*Ursus americanus*) on roads increased in the late 1980s (Harris and Gallagher, 1989). In Britain, the magnitude of road deaths of badgers (*Meles meles*) has been the cause of much concern and may be one of the largest, if not the single largest cause of death (Davies et al., 1987). Likely important implications of road kills for some Australian wildlife are described in Bennett's 1991 review of roads and wildlife conservation. Species mentioned include the threatened eastern barred bandicoot (*Perameles gunnii*), koala (*Phascolarctos cinereus*) and Carnaby's cockatoo (*Calyptorhynchus funereus latirostris*).

WHO IS INTERESTED IN ROAD KILLS?

Road kills appear to be of interest to a wide audience judging from the articles on them in magazines and journals round the world, displays of road kills in museums and cartoons in magazines and newspapers. Many articles have been written on road kills (see for example Brockie (1999), Lalo (1987) and Hill and Hockin (1992)). Museums have displays on road kills (Fig. 6.3). Many meetings and conferences have been held on road-related wildlife mortality (for example, the Transportation-related Wildlife Mortality Seminar; Evink et al., 1996).

For some people, the road mortality of animals is nothing more than part of the cost associated with the benefits of having roads and traffic. For others, the magnitude of road kills is carnage. The deaths of some species arouses much concern. The sight of a koala killed by traffic is bound to cause anguish and indeed the incidence of koala bear road kills has evoked great concern in Victoria, Australia. Contrarily a dead snake on a road or a dead pest species such as the Australian brush-tailed possum in New Zealand (*Trichosorus vulpecula*) is the cause for delight (the possum was introduced to New Zealand and is now a pest of huge proportions). However, road kills in some countries are taken seriously and must be registered with the proper authorities. For example, in the New Forest of southern England, any deer hit by a vehicle has to be reported immediately to the police by the driver.

In the late 1980s, Roger Knutson's concerns over carnage on roads prompted his book *Flattened Fauna; A Field Guide to Common Animals of Roads, Streets and Highways* (1987). At that time, he also announced the formation of a Society, the Simmons Society, for the purpose of gathering information and promoting public understanding of 'road animals'.

Helping toads and frogs to cross roads has become a regular spring activity throughout Europe. Various techniques are employed, among them gathering toads in buckets on one side of the road and literally carrying them to the other. People care about toads. The Duke of Edinburgh wrote a Foreword to a book containing the proceedings of a Road Tunnel Conference held in

130 *Ecological Effects of Roads*

Fig. 6.3. Display at the Museum of New Zealand, Te Papa Tongarewa, Wellington, New Zealand, showing the bones of a possum embedded in the road surface.

Germany in 1989 (*Amphibians and Roads*). In that Foreword he noted: 'Anyone who has had the misfortune to witness what happens when migrating toads cross a road, will appreciate the threat this poses to survival of the species.'

New observations about road kills continue to prompt new research. For example, a new and major research project on barn owls and roads was initiated by the Barn Owl Trust in 1996 after it was found that over half of all reported barn owl deaths in England are road mortalities.

WHAT HAS BEEN DONE TO LESSEN ROAD KILLS?

Driving behaviour, traffic speed and traffic density are factors which contribute to the increase in number of animal road fatalities. Hence some efforts have been made to address driver behaviour and to reduce traffic speeds. Some people advocate defensive driving as a countermeasure against animal accidents. Lower traffic speeds have been introduced in an effort to decrease the number of animals killed on roads. For example, in parts of the New Forest (southern England) the maximum speed is 40 mph (introduced as a means of reducing the number of deer hit by vehicles).

Road signs warning motorists of animal crossings are commonly used and depict many kinds of animals (penguins, eagles, emus, crocodiles, camels and badgers, to name just a few). The effectiveness of these signs has not been well researched. Meanwhile such signs are becoming collector items.

Tunnels, bridges (overpasses and underpasses), fences and hedges have been the main methods employed for lessening road kills. The argument is: either prevent the animal from crossing the road or provide it with a way under or over the road and traffic. In Florida for example, measures have been taken to establish wildlife crossings in the form of underpasses and extended bridges to protect movements of panthers and other mammals (Evink, 1990).

High deer casualties have inspired considerable research. In addition to deer fences, other methods include reduced traffic speeds, devices to alert drivers to animals crossing roads (signs and literature), lighting, mirrors, warning noises (electronic or wind-generated, Fig. 6.4) and reflectors.

Toads, frogs and newts crossing roads as part of their annual migration from wintering site to breeding pond have prompted much discussion about how to reduce mortality of the crossing animals. Closing the road during peak crossing periods and, as already mentioned, collecting the animals in buckets and carrying them across the roads are two methods in use.

Provision of tunnels under roads for amphibians has become well established in many localities in Europe and North America. In Australia, the Queensland Department of Roads (1997) has provided designs for underpasses for roads in the wet tropics. The use of tunnels or underpasses for wildlife is certainly not a new idea, however. The Swiss Highway Department had already installed amphibian tunnels in the district of Cossonay in 1978. Britain's first 'toad tunnel' was installed in 1987. But what would seem to be a simple answer has sometimes necessitated revised designs and incorporation of drift fences or nets to funnel the animals to the tunnel. Even then it has been found that some individuals will approach the tunnel then retreat. The Tunnel Conference held in Rendsburg, Germany in 1989 brought together the most comprehensive research and monitoring of tunnels for amphibians (Langton, 1989). Overall, tunnels in conjunction with drift fences have proven effective for amphibians. However, the design, location and conditions in and around the tunnel entrance also determine the success of these tunnels (see Chap. 4, Fig. 4.7) where tunnels built for amphibians have not been successful

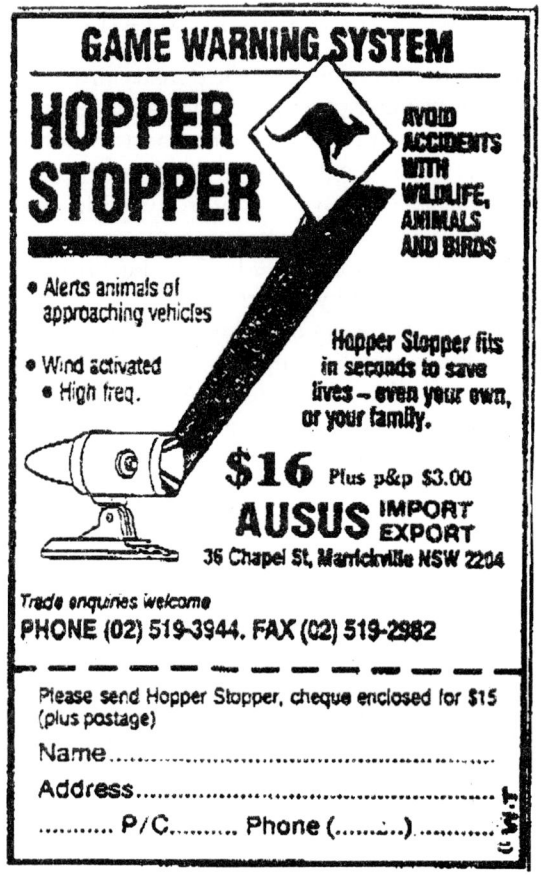

Fig. 6.4. Advertisement for one of the several devices promoted to deter animals from crossing roads in front of vehicles.

this has been due to inadequate fencing ('drift fences') for funnelling the animals to and away from the tunnel and lack of prior planning and study of the animal's behaviour.

Amphibians commonly exhibit hesitation when approaching the entrance to a tunnel, which may well be a response to change in microclimate (small changes in light intensity, humidity and temperature). Intensive, long-term research in Switzerland on amphibian tunnels has led to a standard for their installation. When building pitfall entrances to tunnels, light and dark zones are installed. Animals migrating along a channel pass from a light zone into an interim zone and then a dark zone. Once the animals are in the tunnel, they seem to detect the light at the end of the tunnel and move towards it. The efficiency of light and dark zones in tunnels has been claimed by several authors as good.

Methods to keep deer from roads and to assist in safe crossing have been reviewed in numerous works concerned with deer road mortality and deer crossings. For example, in a questionnaire on mitigation efforts undertaken by state authorities in North America for deer, Romin and Bissonette (1996) found that 95% of the respondents had not conducted scientific evaluations of the techniques employed. The most effective methods for limiting deer crossing roads appear to be fences in conjunction with tunnels and bridges. The height of the fence and its location between road edge and woodland or forest are important for success. Often the fence is at the woodland edge rather than the road edge, preventing animal browsing of grassy strips along the road. It has been suggested that were the fence located at the road edge, thereby allowing the deer to feed in the area between woodland and road, the animals might be less inclined to cross the road.

Some parts of motorways are sometimes well elevated above the surrounding landscape. In such locations, the danger of vehicular collision with birds crossing the motorway is real.

It has been suggested that high hedges might help deflect the flight of birds above traffic flows in locations where road kills of birds on motorways have been persistent. The effectiveness of hedges in deflecting flight birds required careful assessment.

Animal warning devices (whistles and electronic noise-generating devices) attached to vehicles were developed in the USA as early as 1984. A warning device (whistle) marketed by the Save-A-Life Corporation of New York was said to be effective with all animals except cattle, sheep and camels. In Australia, these have been marketed under various brand names including 'Roo-Alert', 'Shu Roo' and 'Hopper Stopper' (see Fig. 6.4).

Research undertaken on the effectiveness of these devices in deterring deer (in the USA) shows no evidence that their ultrasonic frequencies deter animals. In Australia, animal warning devices to alert kangaroos of the presence of vehicles on roads have been developed, namely a whistle audible to kangaroos and an electronic deterrent. Helena Bender's research at the University of Melbourne on the 'Shu Roo' (which claims to be ultrasonic) revealed that only a small portion of ultrasonic frequencies are emitted; when stationary, the signal produced by the Shu Roo is audible to humans (pers. comm.). Her research through behavioural trials with this device on captive Eastern Grey and Red Kangaroos could detect no difference in vigilance, head or ear orientation between control and treatment trials. To date, there seems to be no evidence that whistles or electronic devices deter kangaroos from oncoming traffic. Thus, some animal-warning devices (sonic or ultrasonic) may lack scientific credibility.

So, despite the wide range of methods and some assessment of their effectiveness, further research seems warranted. Road signs and lighting have proven least successful in reducing road kills. It could be that, too many signs warning drivers of animals crossing roads lead to complacency.

Tunnels, bridges, fences, warning signs and traffic controls are methods that deal with the problem after it arises. Prior planning and analysis is by far the better method. In other words, identifying where road kills are likely to occur may enable prevention of the problem in the first place.

Introduction of tunnels etc. does not in itself suffice. The effectiveness of the methods must be monitored and improved. Lastly, objectives need to be defined in terms of road-kill reduction and targeting of localities where levels of road kills are or are likely to be high.

In Summary

- Plan the road and traffic management to avoid or mitigate road kills.
- Introduce tunnels etc. but only after analysis of where best to locate the tunnel and only after the effectiveness of the tunnel design has been demonstrated.
- Reach an accord on how the effectiveness of the various methods proposed should be measured and thereafter monitor the effectiveness.
- Define objectives in reducing road kills and target these localities where road kills do or are apt to occur.

7

Reducing the Adverse Effects

INTRODUCTION

The aim of this chapter is to provide an introduction to the methods for reducing the adverse effects of roads on nature and managing their ecological impacts. The emphasis is on 'introduc-tion' because, in the space available, it is not possible to provide a detailed account of the methods, especially the technical details. Some examples of the technical details have been taken from the literature and the main topics reviewed in the context of road literature. Appendix II presents a guide to ecological impact assessments (EcIAs) and excerpts from three such assess-ments are given in Appendix III. The purpose of these examples is to demonstrate the wide variety with respect to level and detail found in EcIAs. The first is simplistic assessment while the third is a good example of the minimum that should be expected.

THE COSTS

Recent discussions about road costs have sometimes given rise to conflicts over the relative significance of costs of mitigation of adverse effects. For example arguments have centred on the costs of building wildlife bridges in relation to the total costs of the project. Voelk et al. (2001) have published a report on reducing the cost of green bridges. Some argue that this cost is small while others say it constitutes a significant portion of the total costs. In the Netherlands, the total costs of mitigation and compensation have been detailed for several road projects. For example, the costs of offsetting the ecological effects of road A50 between Eindhoven and Oss was 5% of the total construction costs. Arguing the *relative significance* of costs seems futile. The principle of including the *real* costs of projects ought surely to be argued in the first instance.

WHO WILL TAKE NATURE SERIOUSLY?

Until very recently, road engineers and others in civil works departments would not consider the effects of roads and traffic on nature; nature was not taken seriously. The implications of civil works personnel not taking the values of nature into consideration is an important issue that needs to be addressed at many levels. It is because of this failure to take nature seriously that several countries have allowed construction of new roads in wilderness areas (in the last two decades) with total disregard of ecosystems and their values. In the USA no roads are basically allowed in designated wilderness areas and they are greatly discouraged in wilderness quality areas (Marnie Criley, pers. comm.).

Undoubtedly, the challenges of determining the ecological effects of roads are many. This includes deciding what can be achieved in certain time-frames and how the effects can be quantified. Ecological effects can be quantified at certain levels of detail but another approach is to quantify them in a more general manner. This can be achieved by using several criteria (see Chap. 5) or by way of numerical scores for the extent of damage (see chap. 3, Table 3.5). Maps showing the extent and magnitude of adverse effects are also possible but little research has been undertaken on this approach.

The attitude of road authorities has begun to change, and some are taking the ecological effects and impacts on biological diversity into consideration. For example, the Texas Department of Transportation's vision is 'environmentally sensitive' transport systems in which adverse environmental effects (including habitat fragmentation) are addressed.

In the Netherlands, avoidance, mitigation and compensation are three planning concepts designed to address the adverse effects of any project (including roads) on nature (Cuperus et al., 1996, 1999). In Queensland, Australia, the Department of Main Roads (1997) has a manual (prepared by Gutteridge, Haskins and Davey Pty. Ltd.) for roads in the wet tropics ('Planning, design, construction, maintenance and operation best practice manual'). The principles on which this manual is based include the following:

Construction

Undertake construction based on a combination of good forward planning, expert technical knowledge and practical experience with particular reference to best practice environmental management. (The importance of ecologically sustainable development has helped to shape this manual.)

Considerations

Plan to minimise impacts by considering the region's natural and cultural values with particular emphasis on it biodiversity.

Rehabilitate and reinstate areas affected by road projects to a state that is appropriate and consistent with the local habitat and wet tropics regions.

As far as practicable, in all phases of roadworks minimise disturbance of the natural environment.

The need to address the environmental (including ecological effects of roads) is clearly expressed in some general reviews of mitigation (and enhancement); see for example Thrasher, (1983). Most noticeable is the literature from the Netherlands and from Denmark, in terms of both research on development of policy (Bohemen, 1995) and in terms of several publications written for the wider community. Some of these publications include detailed accounts for addressing the problems (for example, the book 'Nature across Motorways', published in 1995 by Rijkswaterstaat (RWS), Dienst Weg-en Waterbouwkunde (DWW), Delft).

The Institution of Highways and Transportation (founded in 1930 and based in the U.K.) is the foremost learned society in the U.K. concerned with the design, construction, maintenance and operation of sustainable transport systems. It produces best practice technical publications included amongst which is The Environmental Management of Highways. Landscape management and road verge management is covered in Chapter seven.

Agencies such as Wallace, McHarg, Roberts and Todd (WMRT) in the USA have produced structured guidelines for inclusion of buffer zones and corridors (see for example Smith, 1993). In New Zealand, Transit New Zealand has produced a manual for cost-benefit analysis of road projects, which includes a section on ecology and some information on mitigation. Some specific mitigation proposals have been drawn up, such as that contained in the Audit of the Future State Highway Number One Route environmental impact report (Parliamentary Commissioner for the Environment, 1990). These mitigation measures include roadside design, drainage and buffers for waterways and habitat disruption.

THE KNOWLEDGE

Much material in the literature deals with both the processes for identifying ecological effects and ways of reducing environmental and ecological effects of roads and traffic. The following topics are included (examples from the literature in Table 7.1):

— planning the route and designing the road with nature in mind;
— reducing pollution (remedying effects);
— buffers and filter strips (remedying the effects);
— barrier tunnels and bridges to reduce road kills (remedying the effects);
— landscaping roadside verges and use of native species (mitigating and compensating the effects);
— mitigation banking (compensation for the effects).

Some guides and manuals have been published on how to assess, avoid, remedy and mitigate the effects of roads and traffic on nature. Far less material has been published on mitigation of other land transport systems such as railways. One such useful report is that by van der Grift and Kuijsters (1998)

Table 7.1: Examples of literature addressing the detrimental environmental and ecological effects of roads (see Reference at the end of the book for complete bibliographical information).

1. Comprehensive literature

Byron 1999. Biodiversity and environmental impact assessment of road schemes.
Byron 2000. Biodiversity and environmental impact assessment: a good practice guide for road schemes.
Erickson et al. 1978. Highways and ecology: impact assessment and mitigation.
OECD 1994. Environmental impact assessment of roads.
Queensland Department of Main Roads 1997. Roads in the wet tropics: planning, design, construction, maintenance and operation.
Ramsay 1994. Roads and nature conservation: guidance on impacts, mitigation and enhancement.
Tsunokawa and Hoban 1997. Roads and the Environment.

2a. Reducing pollution (remedying the effects)

American Association of State Highway and Transportation Officials, Inc. (AASHTO) 1992. Highway Drainage Guidelines (USA).
Atkinson and Cairns 1992. Assessing ecological risks including risks to wetlands.
Isermann 1977. Reduction of contamination and uptake of lead (from exhaust fumes) by plants (Germany).
Kober and Kehler 1987. Mitigating construction impacts on streams (Pennsylvania, USA).
Maestri and Lord 1987. Design and measures for reducing effects of highway storm runoff (USA).
Morgan et al. 1983. Controlling metal leachates and mitigating stream damage (Appalachian Mountains, USA).
Pratt 1984. Design of highway drainage systems (UK).
Stotz 1990. Detention basins to control highway surface runoff (and toxins therein) (Germany).

2b. Use of wetlands as sinks for pollutants (using nature to remedy the effects)

Kadlec 1994. Review of wetland treatment systems (USA).
Reuter et al. 1992. Use of Wetlands for nutrient removal (California, USA).

3. Buffer zones and filters (remedying the effects)

Angold 1997. Buffer zones and oligotrophic communities.
Clinnick 1985. Buffers for protection of streams from sediment (Australia).
Swift 1986. Filter strips to prevent sedimentation of streams (Appalachian Mountains, USA).
Trimble and Sartz. 1957. Logging roads, sediments and streams.

4. Barriers, tunnels and bridges to reduce road kills (remedying the effects)

Evink 1990. Safe crossings for panthers (Florida, USA).
Feldhamer et al. 1986. Fencing and white-tailed deer (Pennsylvania, USA).
Hunt et al. 1987. Tunnels for mammals (New South Wales).
Langton 1989. Tunnels for amphibians (Europe).

Contd.

Table 7.1: (Contd.)

Madsen 1996. Faunal passages and road systems (Denmark).
Mansergh and Scotts 1989. Tunnels for pygmy-possums (Australia).
Murphy and Curatolo 1987. Behaviour of caribou where roads run near pipe lines (Alaska).
Nieuwenhuizen and van Apeldoorn 1995. Mammal use of underpasses (Netherlands).
Owens and James 1991. Brown pelicans and bridges (Texas, USA).
Reed 1981. Mule deer and underpasses (Colorado, USA).
Romin and Bissonette 1996. Deer fences, tunnels and speed controls (USA).
Salvig 1991. Faunal passages and roads (Denmark).
Singer et al. 1985. Underpasses for mountain goats (Montana, USA).
Verboom. 1995. Analytical methods for risks of fauna crossing roads (Netherlands).
Ward 1982 Fencing and mule deer (Wyoming, USA).
Yanes et al. 1995. Vertebrate movement in culverts (Spain).

5. Mitigation banking (compensating for effects)

Howorth 1991. Wetlands (North Carolina, USA).
Lister 1992. Salmon habitat (British Columbia).

---

in the Netherlands. They address in particular habitat fragmentation by railway lines.

The following examples demonstrate the range of published material available (on mitigation of the environmental effects of roads) in terms of depth, specialisation and geographic scope:

1978 *Highways and Ecology: Impact Assessment and Mitigation*. Erickson, P.A., Camagis, G. and Robbins, E.J. Virginia: National Technical Information Services.

1994. *Roads and Nature Conservation. Guidance on impacts, mitigation and enhancement*. Edited by Denise Ramsay. Penney Anderson Associates for English Nature, UK.

1997. *Queensland Department of Main Roads. Planning, design, cons-truction, maintenance and operation of roads*. Prepared by Gutteridge Haskins and Davey Pty. Ltd.

1997. *Roads and the Environment. A handbook*. Edited by Tsunokawa, K. and Hoban, C. for the World Bank. Available on line at the World Bank Web site.

1999. *Biodiversity and environmental impact assessment of road schemes*. Helen Byron, Environmental Policy and Management Group, T.H. Huxley School of Environment, Earth Sciences and Engineering, Imperial College of Science, Technology and Medicine, London.

2000. *Biodiversity Impact. Biodiversity and environmental impact assess-ment: a good practice guide for road schemes*. The RSPB, WWF-UK, English Nature and the Wildlife Trusts, Sandy.

The book by Erickson et al. (1978) draws on experience gained from training courses on ecological impacts of proposed highway improvements sponsored by the US National Highway Institute of the Federal Highway Administration. The authors list relevant legislation for that time and profile road improvements on terrestrial and aquatic habitats and sensitive biological communities. The importance of preserving wetlands is stressed. Mitigation and enhancement methods for terrestrial environments, aquatic environments and wetlands are described in relation to the various stages of road construction; pre-design phase, design phase, construction phase, and operation and maintenance phase. Examples of mitigation and enhancement measures are shown in Table 7.2.

Table 7.2: Examples of mitigation and enhancement measures (from Erickson et al., 1978)

(a) Terrestrial Environment

Project phase	General ecological concerns and/or impacts	Types of mitigation and enhancement measures
Predesign	Loss of ecologically sensitive areas	Careful selection of corridors and alignments
Design	Erosion of construction site	Design erosion control measures
	Loss and/or disruption of wildlife species	Design of alignments, roadbed, culverts, bridges, and barriers.
	Loss or enhancement of wildlife habitat	
Construction	Erosion of construction site	Implementation of erosion control
	Loss and/or disruption of wildlife species	Implementation of design features
	Loss of wildlife habitat	Minimise cutting and stripping Careful timing of construction Careful management of construction site
Operation and Maintenance	Disruption of biota *within right-of-way* Road kill	Careful management of *habitat with within right-of-way*
	Accidental spillage	Implement design and traffic control Features to reduce road kill Implement plans for potential spill

(b) Aquatic Environment

Predesign	Loss of highly productive areas Loss of Oligotrophic water supplies	Careful selection of corridors and alignments

Contd.

Table 7.2: (Contd.)

Design	Changes in water quality	Design of alignments, culverts
	Interruption of fish migration	Design erosion control measures
	Enhanced eutrophication	Specify types of materials
	Changes in hydrology	Specify provisions in construction
Construction	Loss and/or disruption of aquatic species	Implementation of design features
	Loss of aquatic habitat	Minimise encroachment and alteration of aquatic habitats
	Loss of water quality in receiving waters	Implementation of erosion control
		Careful timing of construction
		Careful management of construction camp
Operation and Maintenance	Accidental spillage of potential toxicants and irritants	Implement plans for potential spill
	Runoff terrestrial herbicides	Regulate application of herbicides in right-of-way
	Changes in aquatic productivity due to road runoff	Implement plans for corrective actions
(c) Wetlands		
Predesign	Loss of national wetlands	Careful selection of corridors and alignments
Design	Loss and/or disruption of aquatic and terrestrial species	Design of alignments, culverts and ditches
	Change in area hydrology	Specify types of materials
		Specify provisions in construction contracts
Construction	Loss and/or disruption of wetland habitats	Implementation of design features
	Loss and/or disruption of wetland species	Minimise encroachment and alteration of wetland areas
	Loss and/or disruption of other aquatic and terrestrial species	Implementation of erosion control
		Careful timing of construction
		Careful management of construction camp
Operation and Management	Long-term increase in concentration of potentially toxic materials	Regulate application of herbicides and other chemicals
		Implement plans for potential spill
		Implement plans for corrective actions

The English Nature guide is written with UK roads in mind but some aspects of the publication are applicable to any road. The author refers to the legislative framework and policy framework (for the UK). The guide is divided into several sections centring on habitat loss and habitat fragmentation, hydrological systems, geomorphological and geological issues, and air pollution. Each section contains information about mitigation and avoidance of effects. Also considered are ways of reducing indirect impacts and methods for ecological restoration.

The *World Bank Handbook* (Tsunokawa and Hoban, 1997) is a detailed and extensive template that provides accounts of practical methods designed to employ environmental impacts assessment (EIA) for planning, constructing and maintaining roads. It is divided into two main sections and employing the EIA approach, commences with justification for environmental assessments of road projects. The EIA process, in the context of road planning and construction, is detailed in section one. This is followed by a section on environmental impacts, their mitigation and their economic valuation. Separate chapters detail the impacts on and mitigation procedures for soils, water resources, air quality, the natural environment, human communities and their economic activity, land acquisition, indigenous peoples, cultural heritage, aesthetics and landscapes, noise, and human health and safety. The last chapter deals with economic valuation of the impacts of road projects on the environment.

The Manual for the Queensland Department of Main Roads is the property of Gutteridge Haskins and Davey Pty. Ltd. and is a best practice manual. It assists in the implementation of best practice in the development of roads within the wet tropics region of northern Queensland, Australia. It has taken into account World Heritage and other conservation values. This manual is a very comprehensive, practical document. In many ways, it is unique in promoting an interdisciplinary approach that combines planning, engineering and ecology. It combines theory with practical examples and is well illustrated with good examples of mitigation. Overall the manual supports best practical guidelines for planning, design, construction, main-tenance and operation (Guy Chester, pers. comm.).

Helen Byron, in her two publications, provides a brief introduction to biological diversity, what it is and how it is being approached in the UK. Her publications are particularly useful in the UK and European legislative context. She also draws attention to some weaknesses with respect to biological diversity and EIAs of roads. These weaknesses include lack of proper use of biological diversity terminology, lack of proper consideration of sites not designated for protection, lack of consideration for non-protected species and lack of consideration of all levels of biological diversity. Most of the material in her publications are in the form of guidance with key objectives of ensuring that roads do not significantly reduce biological diversity at any level but, rather, enhance it. She discusses scoping, baseline conditions, impact prediction and assessment, mitigation and enhancement, EIA preparation, review and finally project monitoring.

WHO HAS AN INTEREST?

Who has an interest in new road projects and is it important to know who these people are? Different terms have been used to refer to those people with an interest in a project. In some countries, the term 'player' or 'stakeholder' has been used as an alternative to 'interested party'. The term stakeholder is defined in some dictionaries as being the third party with whom money wagered is deposited. Carroll (1995) defines stakeholder as 'the individuals or groups who can effect or who are affected by the actions, decisions, policies or goals of the proposed development'. I use the term 'interested parties' in the same sense as Carrol for stakeholder.

Wherever there are threats to changes in land use and wherever there are alternative uses of land, there are potential conflicts. Roads, by their very nature, cut across landscapes in which there are many kinds of land uses. The land taken up for roads competes for land being used for other purposes. When planning and constructing a road it is important therefore to establish, as soon as possible, who should be involved and who the interested parties might be. As a first step, it is useful to conceptualise (Fig. 7.1) who the interested parties might be and how they relate to each other and what their interests might be. The issues range from social and cultural to ecological and hydrogel.

In the UK, there are some organisations that have to be consulted when undertaking EIAs. With respect to roads and over and above specific consultation that must take place with respect to each road project, it is recommended that there be ongoing liaison between interested parties. For road projects, County Naturalist Trusts, local government and government conservation bodies such as English Nature should seek to establish formal and ongoing liaison with the road planning and construction bodies. Such liaison should include reviews of policies and scheduled sites so that effects are minimised. As part of the baseline ecological information, County Trusts could maintain records of significant road verge sites in association with information held by the UK Biological Records Centre.

The benefits of liaison and consultation are easy to recognise but achieving successful liaison is problematical. How to establish a collaborative style of public consultation (so as to reach agreement) and how best to present information? Collaborative planning, which brings interested parties together in an authentic dialogue, is not easy to achieve but can be facilitated by those with the appropriate skills. A valuable reference for this kind is Susskind et al. (1999).

Co-operation between interested parties from start to finish of a project has on many occasions proved to be a worthwhile investment in terms of peoples' time, expertise and money. It is one thing to identify who the interested parties

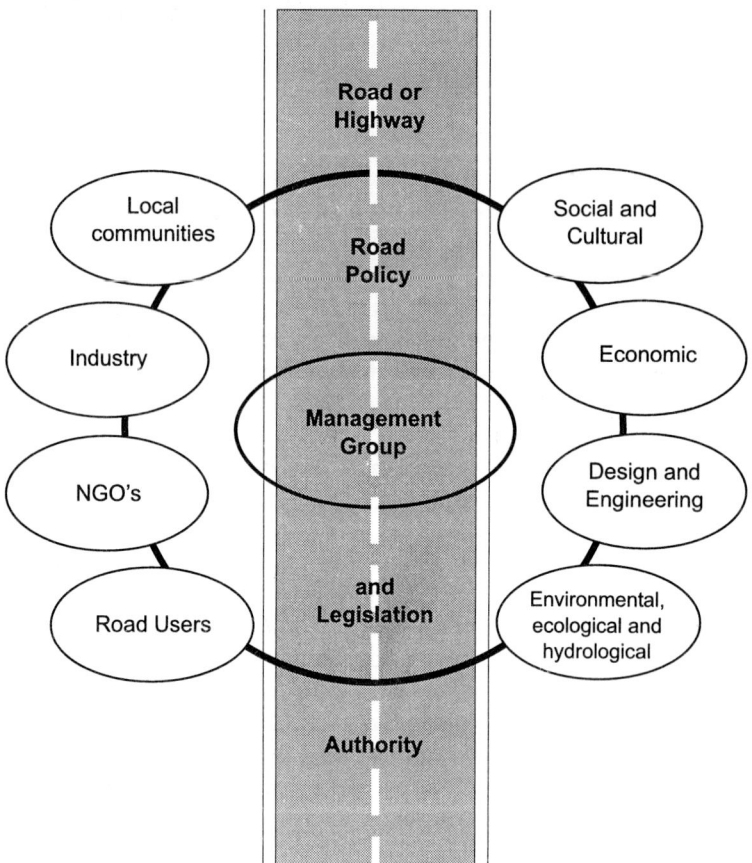

Fig. 7.1. Establishing working relationships and authentic dialogue. A conceptual plan showing the relationships between groups and disciplines underlying management for planning and construction of roads.

are and quite another to extend liaison between groups to a stage where there is management of co-operation between among the interested parties. Not surprisingly, project management committees may need to be established to ensure co-operation between groups of interested parties.

POLICIES, REGULATIONS AND LEGISLATION

The terms 'avoiding' and 'reducing' are used in a general sense here. In the legislation relevant to effects on the environment, these terms have been extended to become more comprehensive. For example, some legislation refers to avoiding, remedying (treat it or rectify it), mitigating (make less severe or alleviate) and compensating for the effects of roads and traffic on nature.

The terms avoiding, remedying and mitigation occur in the New Zealand Resource Management Act 1991. The Purpose and Principles of that Act include the following:

1. The purpose of this Act is to promote the sustainable management of natural and physical resources.

2. In this Act, 'sustainable management' means managing the use, development, and protection of natural and physical resources in a way or at a rate that which enables people and communities to provide for their social, economic, and cultural well-being and for their health and safety while—

—sustaining the potential of natural and physical resources (excluding minerals) to meet the reasonably foreseeable needs of future generations;

—safeguarding the life-supporting capacity of air, water, soil and ecosystems;

—avoiding, remedying, or mitigating any adverse effects of activities on the environment.

There is an argument (supported in Europe) that a distinction should be made between mitigation and compensation and enhancement, in the sense of compensating for land taken for development by setting aside other land (equal in area and quality) for conservation and where possible making habitats better than they were (enhancement).

In the Netherlands, Cuperus et al. (1999) apply the terms 'avoid, mitigate and compensate'. They have adopted and abbreviated the concept 'mitigate' from the American usage in the Environmental Policy Act of 1969 (NEPA, 1970) and use it in the sense of 'minimising, rectifying and reducing effects'. Compensation refers to various kinds of land set aside for nature to balance the area of land taken up for developments.

Reaching a consensus about environment management options can be very difficult. In the context of environmental risk management, Simon Gerrard (1995) refers to the technique 'best practical environmental option, (BPEO) as that which allows the best scientific and ethical weighting of systems based on asking particular groups for their judgements, set in the context of relative costs. The concept of BPEO has a role to play in planning and, according to a Royal Commission on Environmental Pollution, UK (HMSO, 1994), can help in the development of transport policies and in finding solutions to transport problems. BPEO, defined in an earlier Royal Commission report as 'the outcome of a systematic consultative and decision-making procedure that emphasises the protection and conservation of the environ-ment across land, air and water'.

Seven steps provide a framework for the principles of BPEO as applied to transport in the 1994 Royal Commission report on Environmental Pollution (Box 7.1).

Box 7.1:

Principles of BPEO as applied to transport (from HMSO, 1994). Royal Commission on Environmental Pollution, 18th report, *Transport and the Environment*. Reproduced with kind permission of Her Majesty's Sta-tionery Office, permission ref. No. 20001688.

Step 1: Define the Objective

The objective must be defined in terms which do not prejudge the means by which it is to be achieved. It would therefore be inappropriate to set as the objective 'to improve car access to the town centre' or ' to exclude cars from the town centre.' For this illustration the objective might be: **to make the town centre a more attractive and accessible place.** At the same time as the objectives are set, a statement of the constraints imposed on decision-making must be formulated, whether the constraints are legal, technical, social or economic. The implications of alternative proposals should also be considered. Including the 'do nothing' option.

Step 2: Generate Options

All feasible ways of achieving the objective should be identified and the aim should be to find those which are both practicable and environmentally acceptable. In this context, 'practicable' implies that the option must be in accordance with current technical knowledge and must not have disproportionate financial implications. The search should be as wide-ranging and imaginative as possible and should, in this context, include non-transport options. As in this illustration, complex objectives are likely to call for combinations of options to achieve them. The options in this case are likely to include:

— land-use plans to situate facilities where they do not require car access and to discourage developments which are accessible only by car;
— improvements to public transport and to cycling and pedestrian infrastructure;
— car and heavy goods vehicle restraint;
— measures to stagger the such hour (e.g. different starting times for schools and other destinations susceptible to local authority influence, greater use of teleworking);
— development of delivery services for shopping (and perhaps placing orders by telephone or computer);
— encouragement of entertainment and cultural facilities and perhaps residential development in the town centre in order to enchance its liveliness out of business hours;
— construction of new infrastructure.

Contd.

> **Box 7.1 Contd.**
>
> **Step 3: Evaluate the Options**
>
> The advantages and disadvantages of each option for the environment should be evaluated, using both quantitative and qualitative methods as appropriate.
>
> Some effects (for example on atmospheric emissions, noise generation, water pollution (e.g. from runoff), accidents) should be capable of fairly accurate estimation. Others will be less readily quantified (e.g. to what extent the attractive-ness of the town centre has been increased, how sympathetic any new development is) but even here there is increasing evidence which would permit some assessment of the effectiveness of options (e.g. how much a pedestrian scheme might enhance business opportunities).
>
> **Step 4: Summarise and Present the Evaluation**
>
> The results of the evaluation should be presented concisely and objectively, and in a format which can highlight the advantages and disadvantages of each option. The results of different measurements and forecasts should not be combined if to do so would obscure information which is important to the decision. The cumulative effects of a series of complementary options should, however, be spelled out.
>
> **Step 5: Select the Preferred Option(s)**
>
> The choice will depend on the weight given to the environmental impacts and associated risks, and to the costs involved. Decision-makers should be able to demonstrate that the preferred option(s) do (es) not involve unacceptable consequences for the environment.
>
> **Step 6: Review the Preferred Option(s)**
>
> The proposed option(s) must be scrutinised closely to ensure that no pollution risk (including regional and global ones) has been overlooked. It is good practice to have the scrutiny made by individuals who are independent of the original team.
>
> **Step 7: Implement and Monitor**
>
> The achieved performance should be monitored against the desired targets, especially those for environmental quality. This is intended to establish whether the original assumptions were correct and to provide feedback for future developments.

— define the objective,
— generate options,
— evaluate options,
— summarise and present the evaluation,

- select the preferred option(s),
- review the preferred option(s),
- implement and monitor.

Almost as an afterthought, the Royal Commission report notes that public involvement and consultation should be an integral part of the decision-making (especially in steps 1,2,4 and 5).

Generating options seems to be an easy step but the range of options naturally depends on the knowledge, skills and values of those involved. The range of options generated could therefore be limited by the qualifications of the contributors. But who are the options for? Are they for the current generation or for future generations. Are they for the local community or for tourists or commuters? Evaluating the options (step 3) is probably the most complex and challenging step. What methods could be used to evaluate? What criteria to use? Are environmental costs considered. BPEO may have a role but perhaps only as an overall structure on which to explore options and solutions.

Roads and road traffic have often been targeted with respect to regulation. Perhaps not surprisingly, some chagrin has been expressed about the increasing number of regulations regarding impacts and maintenance of roads and also the costs of compliance (see for example Tarrer et al., 1995). There are regulations specific to environmental effects of roads and regulations noteworthy with respect to identifying and dealing with impacts of construction such as roads. Perhaps the range and number of regulations is an indication of the magnitude and range of environmental effects of roads, some of which lead to cumulative and long-lasting effects.

Road projects have immediate effects on the environment as well as medium term and long term. The importance of recognising both the spatial and temporal effects of roads was noted in a report from the Organisation for Economic Co-operation and Development (OECD) on *Environmental Impact Assessment of Roads* which appeared in 1994. The report reviewed the planning process in OECD countries and the strategic aspects relevant to environmental requirements (OECD, 1997). As a follow-up, an OECD seminar was held in 1994 for the purpose of developing a series of recommendations (Box 7.2). The environmental recommendations proposed by the Monte Pellegrino Seminar were very human orientated and impelled by sustainable development.

In the UK, construction of motorways and express roads requires an environmental assessment under the EC Directive 85/337 on environmental assessment, while such assessments are discretionary for other roads. However, it is likely that local authorities will require an environmental assess-ment whenever the road is 'likely to have a significant effect on the environ-ment by virtue of its nature, size or location' (Article two of the Directive). In the UK report from the 1994 Royal Commission on 'Environmental Pollution', there is a strong recommendation that all road construction proposals be subjected to environmental assessment. However,

Box 7.2:

The ten recommendations of Monte Pellegrino (from OECD. 1997. *Road Transport Research Outlook 2000*).

I. OECD member countries should examine the scope for integrating Strategic Environmental Impact Assessment (SEIA) into the overall assessment of transport policies and plans at various levels (e.g. conurbation, regional, national, international). All transport modes and vehicles must be taken into account in order to improve the balance and complementarity between modes and achieve sound and sustainable mobility.

II. The SEIA, for both national land-use planning and transport management, should be co-ordinated as much as possible and should serve as a guide and reference to Environmental Impact Assessment (EIA) of projects.

III. Environment must be considered by decision-makers as important a factor as mobility, accessibility, safety and economics. The use of well-defined goals and policies is essential for formulating alternative actions. New emphasis and new methods are needed to actively seek better environmental conditions, instead of only avoiding or mitigating environmental damage.

IV. Efficient co-ordination between the administrations responsible is essential for the success of the work carried out by road agencies. Decisions by transport authorities should be based on a joint process shared with the other agencies involved; they must work together to determine purpose and needs, scope of impact assessment and the environmental mitigation and improvement measures.

V. SEIA and EIA must integrate cultural, social and natural background. Road administrators must consider social values and keep in mind new ethics, such as those resulting from the Rio Agreement: the precautionary principle and the ability of future generations to meet their own needs.

VI. SEIA and EIA methodologies should be part of the educational curriculum of road administrators, in order to enhance their expertise in project and policy formulation.

VII. Short- medium- and long-term effects must be considered. SEIA and EIA should not be single actions but a continuous, long-lasting process. The responsibility of road designers does not end with the project, and the decisions resulting from SEIA and EIA need monitoring, follow-up and review.

VIII. Participation by the public and the other parties of the decision process require involvement at an early stage of the process. Consequently, project leaders need to be trained in the techniques of communication.

IX. Knowledge about short-, medium-, and long-term effects of roads on the environment must be enhanced. Research on monetary and non-monetary

Contd.

> **Box 7.2 Contd.**
>
> valuation should be developed. National policy must determine how these approaches are used in the decision-making process.
>
> X. The highway administrations of member countries should undertake case studies to examine the possibilities for improving SEIA and EIA methods, and then share their experience. Harmonisation of SEIA and EIA methodologies is necessary, especially for international projects. However, constraints of ensuring conformity which could undermine the efficiency of national actions should be avoided. The process of harmonisation and the development of a common language should continue through ongoing international mechanisms.

the commission goes on to say that the fact that such assessments are carried out does not mean that satisfactory standards have necessarily been achieved.

Article 3 and Annexure III (3) of the EC Directive make it clear that both direct and indirect effects of a project should be considered as part of the EIA, including effects on 'human beings, fauna, flora, soil, water, air, climate, any interactions between the foregoing, material assets and the cultural heritage'. Ecological impacts pertaining to the viability, sensitivity and value of ecosystems, habitats and species which might be affected, must be part of the overall assessment process.

In the USA, the environmental assessment process is a cornerstone of the National Environmental Policy Act (NEPA) and provides a framework for assessing the effects of road projects and EIAs. In addition, there has been significant development with the new National Forest System road management strategy. The US Forest Service has 617,600 km of classified Forest System roads under its jurisdiction. The National Forests contain an additional 219,200 km of roads. The Forest Service has embarked on a national transportation policy assessment. Within two years of the effective date of the final road management strategy, each national forest unit must complete a road analysis. According to Marnie Criley at Wildland CPR, this is a step in the right direction. She adds that this is an opportunity for local people to engage in the new planning processes.

In New Zealand, the requirements for EIAs are described in the Resource Management Act (RMA) 1991 (see above) and the contents of an impact assessment are outlined in the Fourth Schedule of the Act. In an article in 'Local Authority Engineering', Brown (1993) describes how the New Zealand RMA has major impacts for road projects, including that of EIAs.

Five years later, however, comments in a report from the New Zealand House of Representatives (1998) raised some concerns about the limitations of the RMA with respect to roads. That report notes: 'the environmental effects of road transport in New Zealand are very significant and are getting worse'. The report goes on to say that the RMA does not deal with effects arising from vehicles on public roads and that the ability of the RMA to control the direct and cumulative effects of roads is unclear. It seems that the RMA is least likely to control the effects of road transport on the State highways. It might be a useful exercise to assess the management of impacts of road building in New Zealand since the enactment of the Resource Management Act.

REDUCING ADVERSE EFFECTS

General methods

Some general methods are shown in a simple pictorial representation (Fig. 7.2) and further details can be found in van Bohemen (1995), Salvig (1991) and Southerland (1995). Case studies on mitigation of the ecological effects of roads are now growing in number. It would be useful for an agency to establish a database of these case studies and make that database widely available. One straightforward and striking case study is that of the M40 (Waterstock to Wendelbury sections) in the UK. This case study gives several examples of good practice (Box 7.3).

Box 7.3

A mitigation case study (from *'Roads and nature Conservation. Guidance on impacts, mitigation and enhancement'*, published by English Nature).

This more recent scheme demonstrates several examples of good practice, and is also a well-considered integrated mitigation package accepted as adequate compensation for habitat loss and fragmentation.

- The original Otmoor Route was altered to avoid damage to a number of SSSIs, and as a response to public pressure.
- An ecological assessment was conducted for the whole new route.
- The route was planned to wind round a sequence of SSSI woodlands which form part of the forest of Bernwood, which lengthened the route at a cost of £1.9 million.
- Objections by two local landowners pushed the road into 2.6 ha of the edge of Shabbington Wood SSSI. The area affected was a conifer plantation but with important black hairstreak (a rare species) butterfly populations on mature blackthorn fringing hedges.
- The major black hairstreak colonies in a fringing hedgerow were avoided, but smaller ones were affected.
- A mitigation package was agreed with EN which involved translocating the affected blackthorn into a 3 ha area, obtained by agreement with the landowner, adjacent to the wood and road. In addition, irregular strips of trees and shrubs interleaved with wildflower patches sown with seed from an adjacent hayfield were established to provide nectar and breeding sites. The blackthorn planted was propagated from suckers of the native stock as the butterfly is so selective of the genetic type of tree. Deer and rabbits were fenced out.

Contd.

> Box 7.3 Contd.
>
> - Systematic monitoring is being carried out and so far brown hairstreak has appeared (a scarce species) which also feeds on blackthorn. Some black hairstreak pupae survived on the transplanted blackthorn, but some plants succumbed to drought. No black hairstreak have been recorded so far (2 years) in the new area, but the habitat is not yet suitable.
> - New hedges have been established, both along the M40 boundary and between woods, to improve their connectivity.
> - New wildflower grassland are being created along the roadsides.
> - Underpasses for agricultural use were also designed to accommodate the passage of deer, with false cuttings to provide a screen bank, and a grass-crete surface. Deer fencing to prevent road deaths was erected where necessary.
> - Badger tunnels were incorporated where needed, together with sufficient fencing.
> - Additional planting was carried out with agreement of concerned parties.
> - CPO orders were used to obtain mitigation for habitat creation.
>
> Schemes are rarely perfect, and a number of measures could have been improved. These include:–
>
> - Having an on-site Clerk of Works conversant with the ecological and landscaping needs so that, for example, damage to hedges by one of the service industries could have been avoided.
> - More extensive CPO areas for the borrow pits to create better ponds and associated habitats afterwards.
> - Less pressure on the contractors to complete early or on time, which led to short cuts and potential damage through water extraction to a nearby SSSI.

CLOSE THE ROAD OR DO NOT BUILD ONE

It would be a significant milestone were there agreement to close and decommission some roads. Roads in relatively undisturbed areas are candidates for closure. Rather than promoting new roads in these areas, proposals to close extant roads should be made. The Forest Service in the USA has proposed decommissioning of 160,000 km of roads.

Of all the ways to avoid ecological impacts, that of not building the road must be the most important and of particular relevance to avoidance of habitat fragmentation and subsequent incremental damage to nature. If the ecological impact assessment reveals adverse effects that cannot easily be avoided, remedied or mitigated, the road should not be built. Roads are not a solution to traffic problems. New roads in wilderness areas must surely be assessed

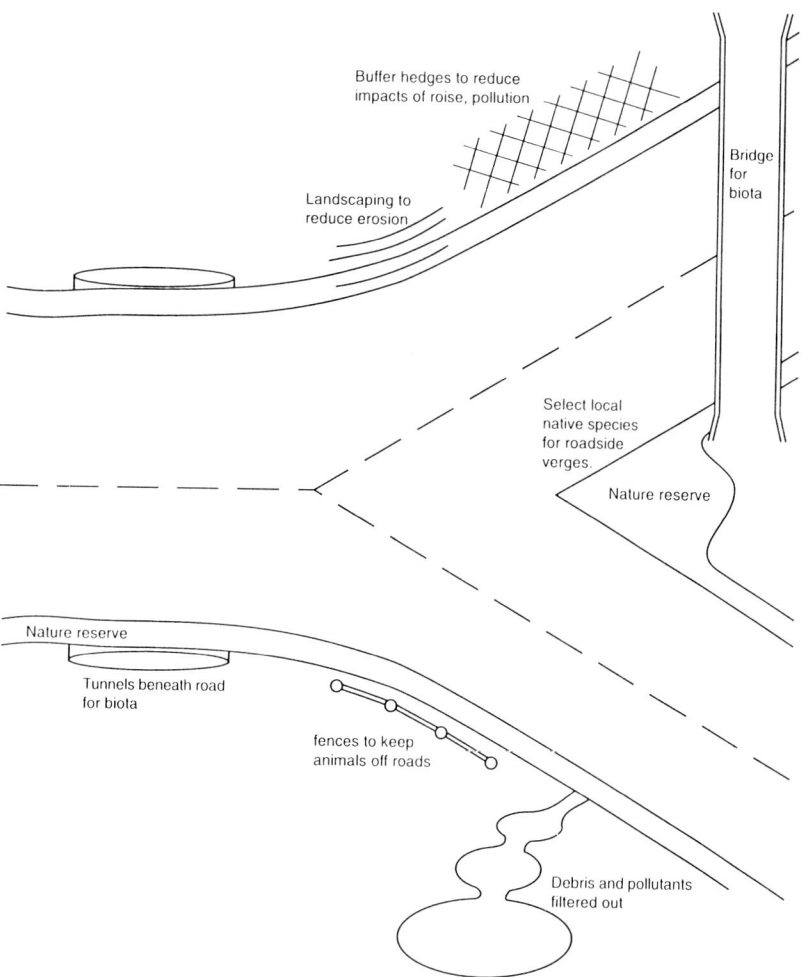

Fig. 7.2. Pictorial representation of the general methods for reducing ecological effects of roads and traffic.

against the extent of ecological damage and the benefits of preserving natural areas.

If the road must be built, alternative routes must be considered. An alternative route might lessen the overall ecological effects. Other forms of transportation such as canals or railways may be alternatives with less short-term and long-term damage.

154 Ecological Effects of Roads

Road design

The design stage affords many opportunities to address the ecological effects of roads. For example, culvert design can be improved to reduce the effects on streams (Kober and Kehler, 1987). Roads cut into the side of a hill are subject to failure planes; several methods are available for addressing this problem. Examples in Fig. 7.3 come from the reference manual *Roadside Bioengineering* by Howell (1999).

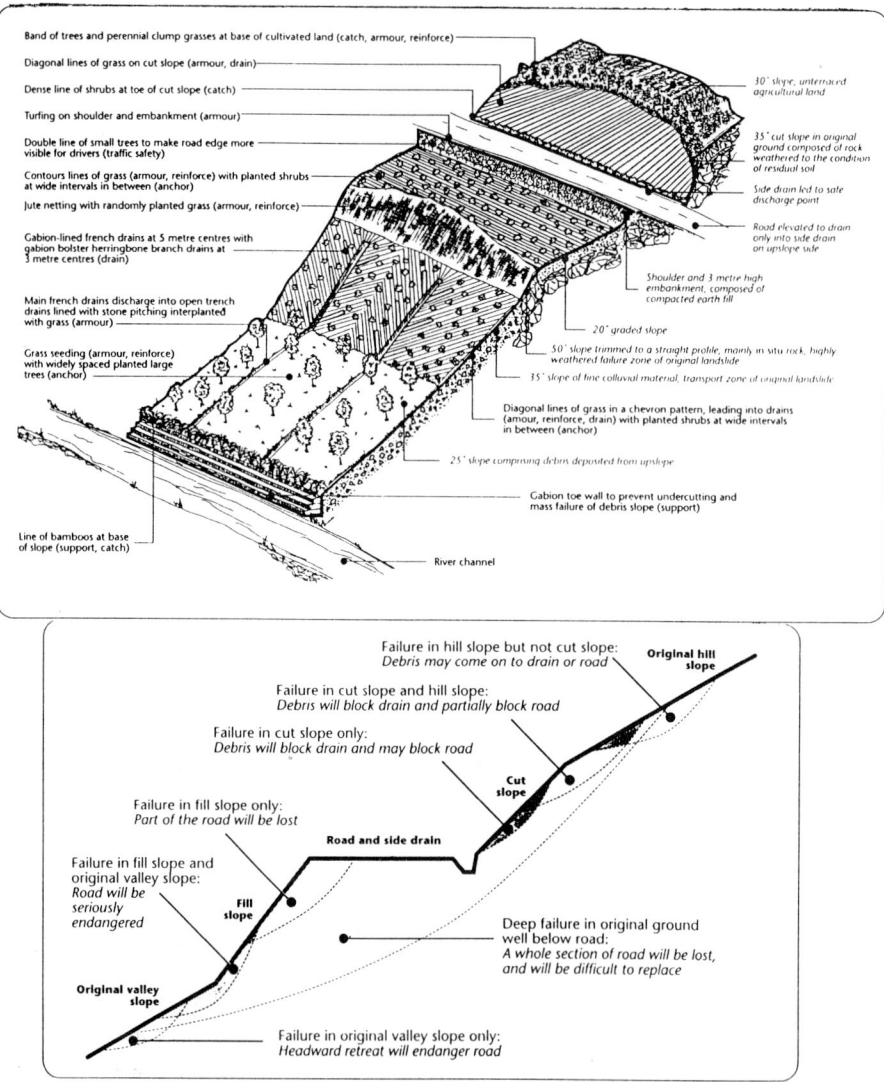

Fig. 7.3. Above Hypothetical example showing methods for stabilising a large slope. Below, Possible failure planes around a road on a slope.

Policies

Identify the existing policy and legislative framework for managing the effects of roads and traffic. What are the policies, who is responsible for what, and what is the relevant legislation? In New Zealand for example, two acts are relevant at different levels (national and regional). The Land Transport Act: National Land Transport Strategy and Regional Land Transport Strategies. The Resource Management Act requires both Regional Policy Statements and District Plans.

Some governments are establishing policies that require consideration of ecological effects of roads together with unfavourable environmental (pollution, noise) and social effects. Policies and planning have an important role to play but so also has the assessment of policies. For example, the Dutch Government has introduced policy instruments for use at a regional level in transportation and in this context a computer model was developed to evaluate the policy measures (Canters and Cuperus, 1997).

Planning and assessments

There are opportunities to address some of the problems of environmental effects of roads before construction (during the planning stages), during construction and after construction. The minimum of what should now be expected is presented in Appendix III (example 3).

Impact assessments and inventories help in assessing the location and route of the road, thereby avoiding or reducing impacts on areas of conservation importance. The effects of the types of materials, including the possible chemical effects of aggregates used for road building, can be assessed before construction. Pollution prevention, habitat enhancement and management, and also mitigation can all be considered at the planning stage.

Appropriate planning, together with the use of GIS for spatial analysis, can help meet this challenge and many models are now available for doing this. In China for example, GIS-based map overlays have been used for the comprehensive assessment of road environmental impacts (Li et al., 1999). Another recent example comes from road revitalisation in the Democratic Republic of the Congo (DRC). A strategic approach to road reconstruction has also been proposed (Wilkie et al., 2000). Spatial models at a national level are presented and the authors suggest mandatory inclusion of the following information:

— road network location,
— location of resources such as grain, coffee and timber,
— centres of internal demand,
— key export points,
— alternative transportation routes such as waterways,
— location and status of protected areas,
— location and value of remanent biological communities.

Although baseline information and spatial analysis expertise is not available in the DRC for such strategic approaches to road planning, much of the information is available on a CD ROM (http://carpe.umd.edu) developed by the Central African Regional Programme for the Environment (Wilkie et al., 2000).

Clearly, an integrative approach is necessary, i.e., one which draws on many disciplinary skills and that uses the powerful tools of GIS. An integrated conservation plan, which helps to minimise environmental impacts that arise from road construction and road widening, has been described by Carr et al. (1998). Similar to the Statewide Greenways Plan developed by the Florida Department of Environmental Protection and Florida Greenways Coordinating Council, the GIS model used by Carr and colleagues helps to identify the interface between conservation values and road construction plans. At a fairly large spatial scale, the approach is useful at several stages, including identification of alternative corridors, design of new road alignments, detailed design of existing road expansions and determination of mitigation strategies. The scale is such that smaller ecologically sensitive areas are not represented. Nevertheless, Carr et al. (1998) conclude that GIS has a role to play in identifying the interface between conservation and road planning because:

— it holds potential for avoiding environmental conflict through early identification of areas of concern;
— where areas of concern arise, an opportunity to prioritise them is provided;
— areas to serve as mitigation can be identified.

Avoiding soil compaction and damage to roots of trees

Several methods have been developed to avoid or minimise the effects of compaction of roots of plants including trees. In the Amazon rain forest, geotextile grids are used to support gravel roadways on soft soils. The Company Maxus reports that by narrowing the road fewer trees are damaged.

In other parts of the world, roads through forests have been elevated to avoid soil compaction around tree roots. In New Zealand for example, Transit New Zealand was awarded the International Road Federation Global Achievement award for work that included bridges over the roots of ancient Kauri trees. Footpaths or sidewalks have also been elevated when the widening of a road has threatened to damage protected tree species (Fig. 7.4).

Ecological restoration and compensation

Habitats, biological communities and ecosystems of significance must be avoided. These can be identified in the planning and assessment stage. Human-made structures can be recreated but nature can't. There may be a case for ecological compensation, however (Cuperus et al., 1996).

Fig. 7.4. Widening of a road entailed compaction of the soil near a group of protected trees. Instead a boardwalk was constructed to serve as a footpath.

Ecological restoration around roads, on road verges and even on traffic islands should be considered (Morris et al., 1994; Sawyer, 2000).

An ecological compensation or nature compensation principle has been incorporated in government policy in the Netherlands. Ecological compensation aims to recover those ecological functions and natural values that still remain affected after maximum effort has been made to reduce the impact of

Fig. 7.4.

Fig. 7.4.

intervention (mitigation). In other words, the aim is to have no net loss of area and quality through implementation of mitigation and compensation measures (Cuperus et al. 1996). Nature Compensation Plans (NCPs) have been undertaken in the Netherlands for roads and Cuperus et al. (1996) have described preliminary methods for deriving ecological compensatory measures.

Conservation banking

The practice of mitigation banking or conservation banking and ecological compensation is becoming more common in connection with road projects. In the USA, mitigation or conservation banking has been introduced as a means by which developers have the opportunity (when their project is approved) to buy areas from reserve land (conservation bank) which is then set against other land taken for development (see for example the Federal Highway Administration 1994 Report No. FHWA-PD-94-004, 'Wetlands and highways: a natural approach'). This is not new idea but one gaining in popularity as a means of mitigating the effects of projects on nature.

Mitigation or conservation banking is simply an extension of compensation for loss or damage to habitats by providing the establishment, or enhancement of habitats for wildlife elsewhere. Two kinds of compensation measures have been suggested by Cuperus et al. 1999. They distinguish between 'in-kind compensation measures' whereby habitat area of the same size and quality is set aside as compensation and 'out-of-kind compensation' whereby any kind of habitat serves as compensation. In the Netherlands, the compensation principle is the policy line adopted by the national Government and is enacted on a voluntary basis.

The monetary costs of combined mitigation and compensation measures have been determined for some road developments. In the Netherlands for example, the costs for mitigating and compensating for ecological effects of highway A50 between Eindhoven and Oss (30 km) was 5% of the total construction costs (Cuperus et al., 1996). For another road construction in the Netherlands, 40 km of A73, 5.8% of construction costs were budgeted for mitigation and compensation (Cuperus et al., 1999).

Fences

In many countries, roads are lined with fences to prevent animal access. The design and location of the fence have to be considered in relation to the behaviour of the animals and the geography of the area. The cost of fencing may prohibit continuous fencing along all main roads. That being the case, criteria have to be established as aids in the decision-making process.

Tunnels, underpasses and wildlife bridges

Roads are barriers for some wildlife and they fragment habitats. Considerable literature is available on tunnels or overpasses (see for example Reed et al., 1975; Reed, 1981; Singer et al., 1985; Travis and Tilsworth, 1986; Hunt et al., 1987; Mansergh and Scotts, 1989; Evink, 1990; Yanes et al., 1995; Madsen, 1996; Clevenger and Waltho, 2000), dealing mainly with their construction and assessments of their effectiveness. Road kills are doubtless the most important effect and engender the most concern. How can road kills be avoided (see also the end of Chap. 6)?

Reflecting lights alongside roads may deter mammals from crossing a road. Whistles and electronic devices are attached to vehicles sold in the USA and Australia which purportedly deter large mammals. There is little evidence to show that they are effective and indeed some could well be a hoax.

Wildlife bridges (ecoducts or wildlife overpasses) may constitute a very notable attempt to reduce the effects of road fragmentation of habitats (Chap. 4, Figs. 4.1 and 4.8). However, little research has been done to determine the effectiveness (in terms of use) and impacts (in terms of population size, fragmentation of population and gene flow) of these attempts to reduce barrier and habitat fragmentation effects.

Traffic rules and traffic regulations

Install signs warning motorists about animals on roads, decrease traffic speeds and introduce traffic regulations.

Buffers

The concept of buffer zones (undisturbed areas or strips) and filter strips (undisturbed except to provide access; Clinnick (1985)) has long been popular in conservation and has been researched with respect to several kinds of impacts, not all of which are relevant to roads. For example, the use of buffer zones to minimise effects of herbicide spray drift was researched by Marrs et al. (1992). Buffer zones on field margins have been developed in Britain with the aim of encouraging habitats for wildlife including beneficial invertebrates. The use of buffer strips to absorb pollutants has been discussed by Angold (1992) and by Curzydlo (1985). It seems that some kinds of dense vegetation may act as sinks for some pollutants.

It may be necessary to have buffer zones between the road and any significant habitats nearby (Angold, 1992; Bickmore, 1990; Clinnick, 1985; Johnson, 1990; Trimble and Sartz, 1957). How wide should a buffer zone be is a common question and one which has often been answered with respect to prevention of sedimentation of streams (see Table 5.1). Very little research has been undertaken on how to identify the optimum widths of buffer strips alongside roads. One way of approaching this is to research the nature and extent of the impacts from roads and traffic on wildlife communities. Angold (1992) has undertaken a very detailed analysis of the effects of roads on heathland communities in southern England.

Landscaping and planting

Landscaping and planting roadside verges to reduce erosion and to provide habitats for wildlife including threatened plant species has been widely researched. (See p. 26)

Role of road verges in conservation

Raise the profile of the value of roadside nature reserves! In Australia for example, there is an established Roadside Conservation Committee of Victoria (Gilbert, 1998). In the USA the concept of 'ecological highways' has been advocated (Evink, 1998). There are designated scenic highways so why not 'ecological highways'?

Don't mow the grass! Or at least change the grass mowing regime to one less frequent and timed to augment the diversity of grasses. All too often, grass verges on the sides of roads are mown because they 'look untidy'. Nature may look untidy but then a well-managed grass verge with mowing regimens less frequent than usual can lead to a beautiful sward of wildflowers. Less mowing saves energy and reduces the costs of roadside management.

Plant indigenous wildflowers and shrubs. The area of roadside verges is such that it is now well recognised that road verges can make a valuable contribution to conservation. They are linear habitats (van Bohemen, 1995). In addition to the verges, the central reservation strips of motorways and also the inner circle of roundabouts (traffic islands) can be used for conservation (Chap. 4, Fig. 4.7). There has been much research on wildflower establishment and management of roadside verges (e.g. Wells and Bayfield, 1990; Schaffers, 2000). Unfortunately, the establishment and management of roadside verges with indigenous plant species is currently overshadowed by purely engineering and landscape considerations. However, there are now manuals, some underpinned by extensive research, which provide details of roadside verge establishment and management from an ecological and conservation perspective (Chap. 3).

Management of grass on verges can be a difficult issue. Unmown or infrequently mown grass may help to increase plant diversity. Long grass may also augment the abundance of small mammals and other small ground-dwelling animals. Where long grass is adjacent to the road, this could increase road kills. Management of grass verges may therefore have to include regular mowing of the first metre or so of grass to discourage small animals from moving onto the adjacent road.

Reducing pollution

The pollution effects from vehicles has led to much action in terms of avoidance and remediation. Many regulations deal specifically with effects on environmental quality such as air quality. In one report (Kammerbauer et al., 1986), reduction of toxic exhaust emissions by means of catalytic converters is advocated to minimise their effects on spruce and other conifers.

Alternative transport systems such as rail or canals would in general reduce the amount of road pollution and require less fossil fuels.

Publications on pollution from vehicles and their effects on wetlands abound. This literature centres mainly on avoiding contamination of wetlands (see for example Kober and Kehler, 1987) and containment of surface runoff from roads. By way of contrast, research has also been done on the use of wetlands as a sink for urban water runoff (for example, Reuter et al., 1992).

Another form of pollution is litter discarded by humans, either from moving vehicles or more commonly left at rest areas or lay-bys. The effectiveness of litter bins has been debated. A more recent practice dispenses with litter bins and signs encouraging roads users to take their litter home.

What else?

In recent years, some transportation authorities have commissioned research on modes/means of reducing the ecological effects of roads and traffic. There is certainly a need not only to be innovative and develop new ideas, but to reassess many of the existing methods in different biogeographical regions.

THE ULTIMATE CHALLENGES

Roads will continue to be built because of their perceived value to the economy. They will also be built because, purportedly, demand is not being met by supply. Cost-benefit analysis of new road developments can be very persuasive but it should always be remembered that such an analysis may be just one component of a decision-making process. Roads and traffic generate wealth but they also contribute to human health problems and cause both environmental and ecological damage. The ultimate challenges are to design roads with economic and social benefits but minimal impact on the environment and to strive for a more sustainable transportation system (sustainable in the sense of environmental sustainability). These are simple observations that are all too often ignored.

8

The Ecology of Roads in the Future

RECOGNISING AND ACCEPTING THE ISSUES

Recognition and acceptance of the issues is needed. At the moment, there is growing recognition that more roads do not solve traffic density problems. But recognition and acceptance of the extent of ecological impacts caused by roads and the seriousness of this for nature have barely begun.

The factors contributing to the real costs of roads and traffic are known. Recognition of these factors is one first step towards addressing this complex issue. Computing the real costs and agreeing on how to pay and who pays is the second step.

It is very clear that roads and vehicles have had and continue to have serious and long-term ecological impacts. Giving ecology equal priority with road engineering and road landscaping would be a third step towards addressing the issues.

There always seems to be a demand for more roads and upgrading existing roads. The main contributory factors are the growing densities in traffic and the driving forces of transport economics. However, it has been shown many times over that more roads do not, in the long run, solve traffic density problems. More roads or upgrading existing roads are short-term measures.

Roads are but one of several transport options. Planning on a large spatial and temporal scale may help to identify cheaper options, one of which is rail. All too often the issues are addressed one at a time because it is simpler to do so. Planning transport routes should not be done in isolation of infrastructure and the many other aspects of urban and regional planning.

Road designs in future will require the skills and expertise not just of engineers and planners, but interdisciplinary people with skills and knowledge of ecological engineering.

VEHICLES, ROAD DESIGN AND ROADS IN FUTURE

Road transportation in its present form is, in general, not sustainable from an environmental point of view. This is mainly because road transport is largely dependent on non-renewable resources and by the fact that environmental effects are not fully included in road costs. Over the next few decades, there will be less reliance on non-renewable energy sources for vehicles and more use of renewable energy. Emissions will change and have less impact on the ecology of roadside verges. There will be slow acceptance that more roads cannot be justified and that existing roads have already contributed to widespread ecological degradation and social damage. Alternative transport media will compete with roads. Fewer roads will be built and there will be declines in traffic densities. Rather than thinking about new roads in wilderness areas, there will be a commitment to close some roads already there.

My visions for the future are as follows. Road design in future will commence with selection of routes with least ecological impacts. Wildlife bridges and tunnels will be located and built to minimum standards. There will be ecological restoration of roadside verges.

For every square metre of road there will be at least the equivalent of land set aside for nature. Roads will become linear nature reserves with hedgerows of native species and a wide swathe on either side as a nature reserve. These linear nature reserves will be habitats for rare and endangered species. Roadside verges will be enjoyed by all and traffic will slow to allow travellers to enjoy roadside nature.

AREAS FOR FUTURE RESEARCH

The magnitude and extent of the ecological effects of roads and vehicles easily justifies research in this area.

i) Costs and benefits

There is growing recognition that the cost of roads has long been underestimated. As with the real costs of transportation, there is much need for research on the real costs of roads. That alone would be merely a theoretical exercise; addressing the real costs in practice would be a real challenge for ecological economists.

ii) Least impact road designs

Ecological engineering is a growing discipline and there are many avenues where it will expand. The design of roads, which have least ecological impacts, is a current research and design challenge.

iii) Pollution

Effects of heavy metals (trace metals), while there is much research showing rates and levels of accumulation of metals in roadside biota, their effects have not been well researched. As early as 1976, Smith reviewed lead contamination of roadside ecosystems and noted that our understanding of the effects on biota was poor. More recent reports of the late 1980s continue to mention that we know little about the chemistry of heavy metal uptake in biota, the accumulative effects and the physiology. In particular, there seems to be some controversy with respect to the effects of heavy metal accumulation on forest trees (Backhaus and Backhaus, 1987).

iv) Long-term effects

Very little research exists on monitoring of ecological effects over time and little research on long-term effects. This could be particularly relevant to the ecology of invasive species and dispersal of those species via roads and road traffic.

v) EIAs

There are many Environmental Impact Assessments of road projects but the level and content of the biological and ecological aspects vary; some leave much to be desired. There could be more research on the nature of ecological impacts of road projects. There is a need for an appraisal of the biology and ecology of EIAs and subsequent ecological monitoring of road projects.

vi) Habitat fragmentation

Many authors recognise that fragmentation of habitats by road is perhaps the most important of the ecological effects. Ecological studies of fragmentation are growing in number but reports which analyse the effects of fragmentation by roads are very few. There is room for research on analytical techniques as well as on modelling the likely effects.

vii) Social impacts

More research on the effects of roads and traffic on communities and amenities is required.

viii) Methods to reduce barrier effects

Much has been said about the use of tunnels and wildlife bridges but little research on the effectiveness of either. Considerable research has been done on the design of tunnels, but far less on the design of wildlife bridges.

ix) Buffer zones

The concept of buffer zone, like wildlife corridors, is widely implemented but like wildlife corridors, the ecological aspects of such zones have not been well researched.

x) Creative conservation

There is much scope for expanding creativity with reference to conservation values of roadside edges, central reservations and roundabouts.

ECOLOGY, ECONOMICS AND POLITICS

Road networks continue to expand round the world, an expansion driven by economic greed and shallow political thinking supported by powerful lobbying of road and road transport supporters. Short-term economic benefits lead to more roads and the long-term environmental costs are left for future generations to deal with. The planning and environmental assessments of roads have little ecological content. Ecology is often an 'afterthought' and ecological assessment and research are done 'on the cheap'. Ecology continues to have a low, simple profile compared to economics. Ecology and environmental management principles should become an integral part of all economic, business and commerce education programmes.

What will it take to change political thinking so that environmental costs become an integral part of planning for roads? Will there be a time when the accepted policy is a no road building programme?—despite the growth in motor vehicle traffic.

More roads will never solve the ever-increasing problems of traffic density. To suggest that more roads will ease the burden of choked motorways and city traffic problems is a clear sign of blinkered and shallow thinking.

Appendix I
Definitions

This book is directed towards not just engineers and biologists. It is intended to be used by anyone with an interest in the subject, no matter their specialised expertise or training. It is therefore in the interests of maintaining good communication that the main ecological terms used in the book are defined. This is not to say that these definitions are the only accepted ones.

It is interesting to note in passing the difference in some terms commonly used in Britain and in North America as noted in Wolfgang Zuckermann's book *End of the Road* published in 1991. The following Table is based on that book.

Term in Britain	Term in North America
articulated lorry	tractor-trailer
carpark	parking lot
carriageway	roadbed
lorry	lorry or truck
motorcar, car	automobile, car
motorway, dual carriageway	freeway, expressway
pavement	sidewalk
petrol	gasoline, gas
public transport	mass transport
road verge	road edge
total road area	road right-of-way

Definitions

Roads are lines of communication between places. The noun 'road' is used here in a very generic manner; from tracks used occasionally by vehicles, through earth or gravel roads and roads in urban environments (streets), to highways and motorways (autobahns, freeways). The term 'road right-of-way' is used in the sense of the whole corridor taken up by the surface used by vehicles together with the verges on either side. Traffic is used largely in the sense of motorised vehicles. However, there is reference here to the effects of humans and other animals carried by the various means of road transport.

Road verges

The road verge (Fig. AI.1) is an important aspect of road ecology. For the purposes of this book the road verge is defined as the strip of land beyond the road surface. That strip of land may or may not have plants on it. The road edge is the limit of the formal rsoad and is the boundary line between the surface used by vehicles and the verge.

Road Terminology

Dual carrigeway

Motorway (freeway/expressway)

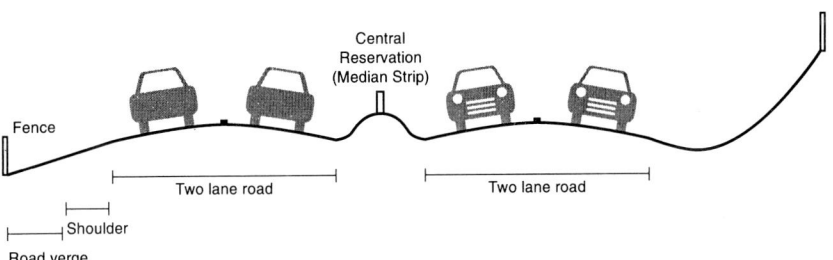

Fig. AI.1: Roads and basic terms.

Linear habitat

The road verge is a linear habitat and in some cases may be an ecotone. In some cases the nature of the road verge may be the same some distance from the road. Alternatively the road verge may be in the form of an ecotone such as grass strips bordering forests. The microclimate of ecotones is diverse and in some localities they have a unique fauna and flora (see also below under habitat).

Wildlife corridors

A term of recent popularity for which a commonly given example is the road verge. Wildlife corridors, greenways, and other similar terms have appeared in much of the literature relating to roads and cities. Any linear landscape feature such as a river bank, a canal, railway verge or road verge has the potential to be a corridor for wildlife, that is, to facilitate movement of wildlife along the linear feature between habitat patches. However, there is not much evidence to show that so-called wildlife corridors are acting as wildlife corridors in the sense of overcoming isolation of populations. They are but linear habitats.

The term wildlife corridor is used here in the sense of a link between habitat patches which facilitates movement of genes between populations.

The term linear habitat is used in the sense of a particular shaped habitat and a conduit for dispersal and dispersion (but not necessarily a corridor).

Ecological effect

The term effect (of roads and vehicles or traffic) is used in preference to impacts. This is because 'impact' is sometimes perceived to mean something physical and something unfavourable. The effects of roads and traffic include impacts (something physical) but also result in other things such as provision of wildlife habitats along road verges. A linear habitat alongside a road could hardly be called an impact.

Environment

Environment means many things to many people. There is a natural component, a built component, social component and living component. Different people identify with different components and may have a narrow view of what the environment is; for example the environment for some may be only their urban-built environment (where they live). Here the term 'environment' is used in a very broad sense and encompasses landscapes, the land, sea, air, water, plants, animals and other organisms and human-built structures.

Biology

Biology is the scientific discipline dealing with the study of life, i.e., plants, animals and other organisms. I include humans with animals. Within biology or overlapping with biology there are other disciplines. Ecology for example is a subset of biology that deals with the interactions between living organisms and their environment. The biology of a road would include information about what is on it and an inventory of the species found alongside it. Ecology would deal with the way the road and traffic affects and interacts with the living organisms. That is, ecology is about physical and natural sequences or systems and interactions. It is also about processes such as movement of energy and nutrients through biological communities.

Wildlife

Whereas in some other books, wildlife is used to refer to only mammals and birds, here wildlife is used to refer to any living organism including plants.

Wildlife therefore refers here to both plants and animals. In the popular media it is not uncommon to hear the expression 'birds and animals'. What should be said is birds and other animals because birds are animals. There is also the mistaken belief that mammals are different to animals. Animals include mammals, birds, reptiles, fish and invertebrates (such as insects, mollusks and crustaceans) and other organisms such as lichens and mosses (Fig. AI.2).

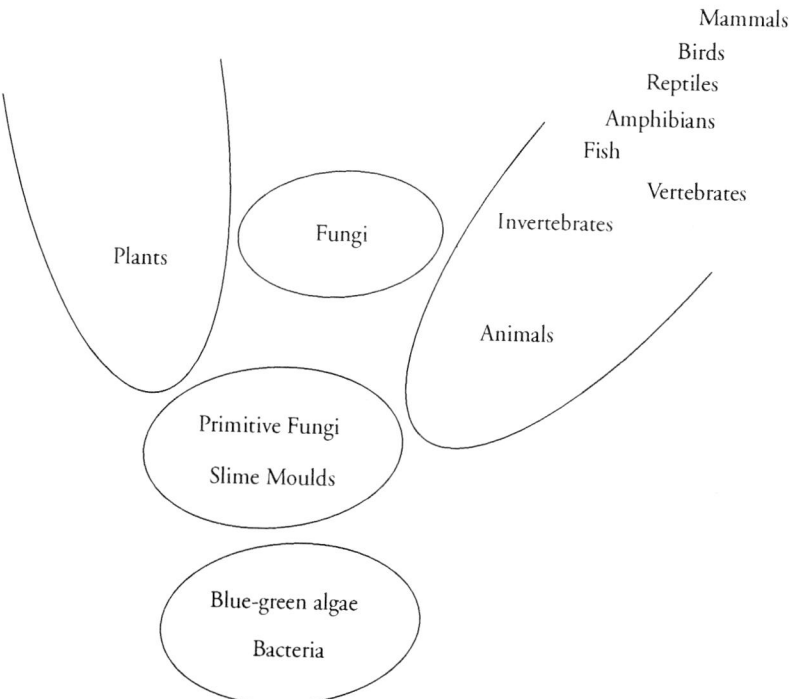

Fig. AI.2. Major groups of organisms.

Biota

In biology there is a conventional way of dividing different kinds of organisms into different groups. Organism could be used for any kind of living being and biota is a term used to refer to all living organisms. Biota are divided into five major groups (Fig. AI.2).

Biological diversity

Wildlife has long been used in biological studies. However, it is noticeable that in more recent publications the term biological diversity starts to replace terms such as wildlife and biota. Biological diversity is defined in the 1992 Convention on Biological Diversity and that definition refers to diversity at different biological levels of organisation. It is a complex term and there are now many different perceptions of what it means. There is no right or wrong definition. As with other technical terms, it is important that the individual state what is meant by it or describe the usage of it. Since biological diversity is so all embracing I suggest that whenever the term is used, it be defined in whatever way it is qualified with respect to the level of biological organisation referred to (molecular? genetic? species or biological community?—to name a few).

Habitat

The place where animals and other organisms live is sometimes called a habitat. The term habitat is a very general one but can be qualified by the predominant physical features of the location or the predominant kind of vegetation. For example, we may speak of alpine habitats, mudflat habitats or lake habitats. Alternatively we may refer to grassland habitats, coral habitats or woodland habitats. The term 'habitat' is widely used in describing impacts on the environment. In an ecological sense the term refers to the environment of a population of a particular species. For example, the ecological literature would refer to the habitat of a particular species of bird or plant. Habitat, as a broad classification of the environment, is more widely used, thus wetland habitats or roadside verge habitat. The term habitat, is used in this book in the wider and more popular context. In general, a habitat can be thought of as having an inner or core area and an edge. A woodland habitat for example ends where the trees end and the zone between the woodland and the adjoining area is known as the edge or ecotone.

Ecotone

Technically an ecotone is a transition between two or more diverse communities as, for example, between forest and grassland or between a soft bottom marine and hard bottom marine community (Odum, 1959). The ecotone is a zone where there are rapid changes in both plants and physical features such as light levels, temperature levels and humidity. By way of contrast, the core area of the habitat is more uniform. In general, the kinds of organisms living in the core area would not be able to live in the ecotone area and *vice versa*.

Biological community

In biology the term biological community refers to a group of organisms that live together. We may envision deep sea communities, river communities, mudflat communities or woodland communities. The habitat is where the organisms live and the biology or biotic community is the assemblage of plants and animals living together. The terms habitat and community are general, widely used terms. They can be used in very specific ways and indeed in ecology, different types of communities are identified and classified often on the basis of the typical mix or type of assemblage of plants and animals found living together. Thus, we can think of woodland communities or wetland communities but in both cases there are many kinds of woodlands as well as wetlands.

Ecosystem

An ecosystem or ecological system is a concept referring to processes or systems and the biological and physical components of the environment. Processes or systems include energy flows (from the sun through plants to various groups of organisms), water cycles, carbon cycles etc.

Ecosystems consist of the organisms or biota, the physical environment (air, water, soil) and processes. We sometimes speak of the properties of ecosystems, referring to the fact that ecosystems provide us with goods and services or act as sources and sinks. We use ecosystems for sources of fresh water, timber and oxygen. We also put our waste into ecosystems and thereby use them as a sink.

Compared to a biological community, it is not possible to draw a line around an ecosystem because it actually has no boundary. The emphasis is on processes not on entities.

Biogeography

The study of the geographic distribution of plants and animals and the reasons leading to those distribution patterns is called biogeography. In conservation studies there are different levels of addressing the issues. Applied aspects of conservation are usually directed towards species or their populations. In the next step of organisation, conservation is directed towards habitats.

Appendix II
Environmental and Ecological Impact Assessments

AII.1 ENVIRONMENTAL IMPACT ASSESSMENTS, THEORY

A systematic and formalised process established long back (dating to the 1960s) to assist in the identification of likely effects of a development on the environment is the Environmental Impact Assessment (EIA). The EIA provides a structured process for identifying likely impacts and also a basis for identifying ways of avoiding, remedying, mitigating and compensating for those impacts. A simplified conceptual outline for an EIA process is shown in Fig. AII.1 and an exemplary contents list for an environmental assessment manual given in Table AII.1.

One particularly important and challenging part of the EIA process is the section on interactions, that is, predicting, describing and quantifying the likely nature and magnitude of impacts on the environment. Should all impacts be identified and should the magnitude of all impacts be assessed as fully as possible? If the answer to the first question is yes, then how can anyone be sure that all impacts have indeed been identified? How can anyone be certain that the effects on the environment will be investigated in as thorough a manner as possible? In some EIAs, attempts to identify the range of impacts may be superficial. However, various methods have been developed to ensure that an EIA is as comprehensive as possible. Such methods include checklists, flow diagrams and matrices of various kinds. All these aid in identification and quantification of impacts. Examples of the use of checklists and matrices in connection with EIAs can be found in any of the many books on EIAs (for example Gilpin, 1995; Treweek, 1999).

Not all environmental effects on the environment are direct effects. Some are secondary and some tertiary. Some effects are cumulative and some synergistic. Flow diagrams help to identify and communicate the secondary and tertiary effects. For example, what environmental effects could a new road development (near an estuary) have on the marine environment? The following simplified chain of events is the first step in constructing a flow diagram.

New roads—increase in area of impervious surfaces—increased flows of fresh water run-off to the estuary—erratic, local changes in levels of salinity—effects on the distribution of bottom-dwelling estuarine species (especially those sensitive to changes in salinity).

Appendix II 175

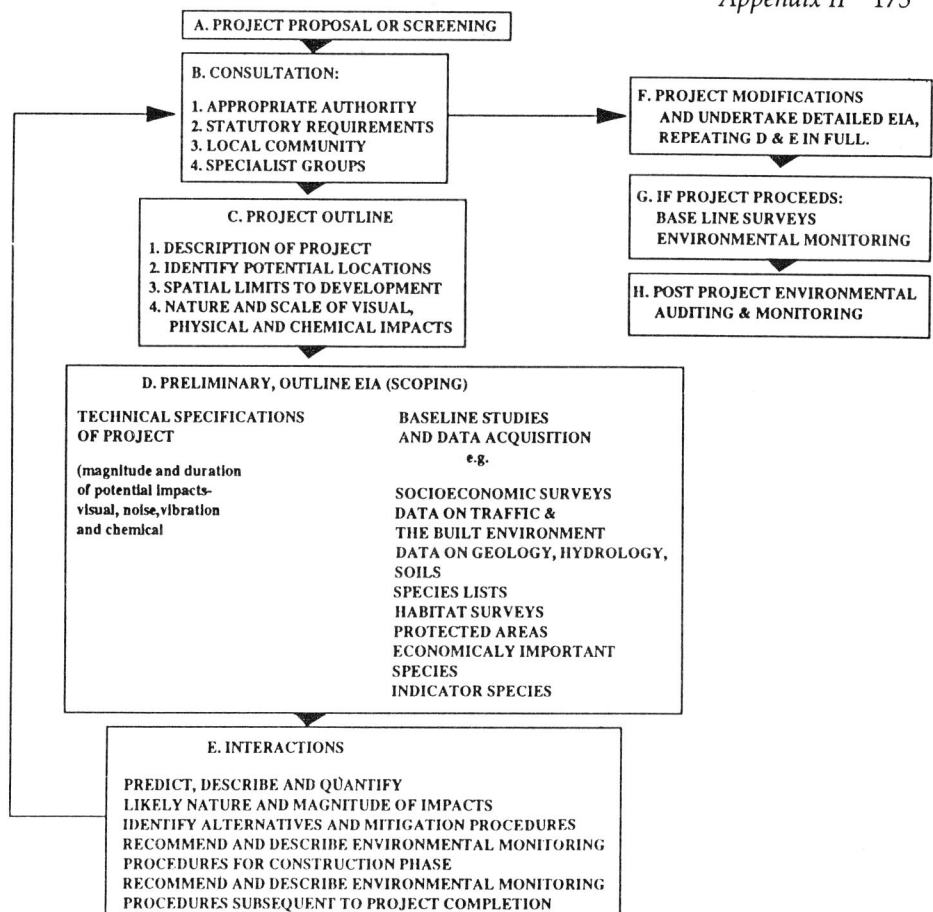

Fig. AII.1. Conceptual framework for an EIA (after Spellerberg, 1991).

The fact that there may be secondary and tertiary effects has to be addressed by the concerned parties and one issue to be agreed upon is: how far to take the assessment of environmental effects in the sense of natural systems. When assessing the likely effects of a road, should the effects on the environment include the effects of the foundry used to make the steel for the vehicles to transport the aggregate for the base of the road? It is impossible to be totally comprehensive. What is important is that the spatial and temporal boundaries for assessing impacts are agreed upon.

The EIA process has long been shown to be a good basis for identifying and minimising the effects of a development on the environment. It provides a structured process. However, an EIA does not necessarily ensure that negative effects are avoided, remedied or mitigated. Nor does an EIA ensure monitoring of the efforts to avoid, remedy, mitigate and compensate for these negative impacts. That such efforts will be undertaken can only result from agreement by the parties involved.

Table AII.1: Suggested list of contents for an environmental assessment manual (from HMSO, 1994: *Assessing Environmental Impacts of Road Schemes*). With kind permission of her Majesty's Stationery Office, permission ref. No. 20001688.

ENVIRONMENTAL ASSESSMENT
What is environmental assessment?
The objectives of environmental assessment
Strategic environmental assessment
Assessment Reports and the Environmental Statement
Assessing 'relevance' and 'significance'

UNDERSTANDING AND ENVIRONMENTAL ASSESSMENT
The nature and scope of matters and effects to be considered
Climate and air quality
Effects on soil and water
Effects on flora and fauna
Landscape
Effects on people as individuals
Effects on communities
Effects on material assets and cultural heritage
Avoidance, reinstatement, mitigation and amelioration
Strategic and long-term effects
Interactions between the above factors
Time scales and discounting

EFFECTS ON POLICY
Policies for conservation
Policies for Environmental Improvement
Development policies
Transport policies
The wider spectrum – environmental policies

TECHNIQUES OF ASSESSMENT
Air pollution
Water pollution
Ecological impact
Visual impact
Land – take
Effects on open space
Effects on agriculture
Traffic noise
Personal stress and other effects on individuals
Disruption due to construction
View from the road
Community severance
Other effects on communities
Heritage and Conservation areas
Vibration

Contd.

Table AII.1: (Contd.)

PRESENTING THE RESULTS OF ASSESSMENTS

Organising the presentation of assessments and the Environmental Statement
The Environmental Assessment Report at Public Consultation Stage – assessing the options
Preparing descriptions:

- The proposal
- The corridor
- The site
- Alternative development proposals considered
- Aspects of the environment likely to be affected
- The effects on natural resources
- Forecasting methods
- Measures to eliminate, reduce or offset environmental effects
- Treatment of time scales

The non-technical summary
Other matters

GUIDE LIST OF POSSIBLE ENVIRONMENTAL EFFECTS
APPENDICES

Environmental Impact Assessment guidelines for roads are not new. In addition to the World Bank guidelines, other international organisations such as the International Union for the Conservation of Nature (IUCN, 1996) have published documents such as 'Tourism, ecotourism and protected areas' in which the question of assessments of new roads with respect to environmental impacts is outlined. The theory of EIA with reference to roads was explored in 1989 in an FAO Conservation Guide ('Watershed Management Field manual. Road design and construction in sensitive watersheds').

In addition to world agency documents dealing with environmental effects of roads, some literature is country specific. For example, in the USA for the state of Washington, Horner and Mar (1983, 1985) reported a protocol for assessing the impacts of road operations on aquatic ecosystems. That protocol offers opportunities to forecast potential aquatic impacts at an early stage of development.

Across the Atlantic, Hodgen and Ford (1985) described planning and design of roads in the UK for what are known as Areas of Outstanding Natural Beauty (AONB). Also in the UK, the Department of Transport (DOT, 1992) has published a review 'Assessing the Environmental Impact of Road Schemes' which includes effects on wildlife. Broad categories in EIAs for roads in the UK now commonly include the following:

- Air pollution
- Effects on farming and recreational activities
- Social effects (considered to be particularly important)
- Noise effects

178 *Ecological Effects of Roads*

- Effects on nature including habitat fragmentation and alteration and also effects on hydrology
- Landscape effects
- Archaeological effects (becoming increasingly important).

Within the context of EIAs, many methods have been proposed for identifying effects (primary, secondary and tertiary) and some have been developed specifically for roads. For example, Lelievre and Serodes (1995) have suggested a cause-effect network. This type of identification is based on three components: actions undertaken, environmental characteristics and stages of the project (Fig. AII.2).

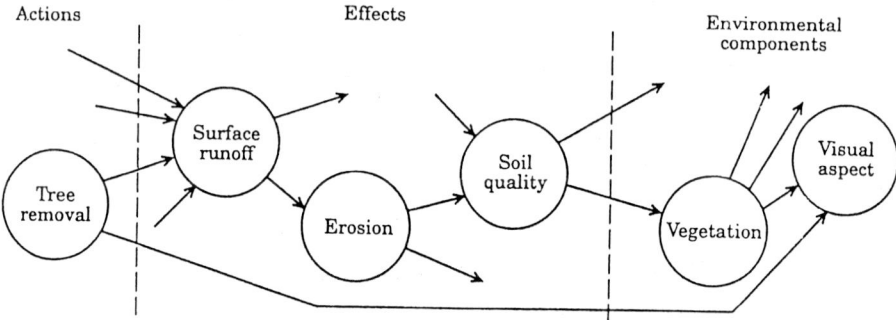

Fig. AII.2. Interconnected elements in a network (from Lelievre and Serodes, 1995).

An EIA is multidisciplinary and may include several components with a disciplinary basis. For example, there may be social impact assessments or geological impact assessments and ecological impact assessments. While there may often be comprehensive EIAs with respect to many disciplines, some EIAs focus on social or geological impacts. EIAs with particular reference to assessment of sediment load and water runoff from potential roads, for example, have been examined by several authors (e.g. Younkin and Connelly, 1981; Horner and Mar, 1983; Kerri et al., 1985; Lord, 1987; Shelly et al., 1987; Smith and Lord, 1990).

AII.2 ENVIRONMENTAL IMPACT ASSESSMENTS, PRACTICE

Despite the many thousands of EIAs that have now been completed in many countries, the biological and ecological components of them were not well examined (Treweek et al., 1993; Spellerberg, 1994). Indeed, for many years (1960s to 1980s) the impacts on living environment were largely ignored or superficially treated and even today there seems to be no agreed standard methodology for ecological impact assessments. Research by Treweek et al. (1993) supports the claim that many EIAs of road projects fail to provide the data necessary to predict potential ecological effects and very few attempt to quantify them.

In 1984, Beanlands and Duinker were especially critical of the sporadic applications of ecological principles in impact assessments. Based on a review of 30 EIAs prepared in Canada, they drew up a number of recommendations:

— identify early on, an initial set of valued ecosystem components to provide a focus for subsequent activities;
— define a context within which the significance of changes in the valued ecosystem components can be determined;
— show clear temporal and spatial contexts for the study and analysis of expected changes in the valued ecosystem components;
— develop an explicit strategy for investigating the interactions between the project and each valued ecosystem component, and demonstrate how the strategy is to be used to co-ordinate the individual studies undertaken;
— state impact predictions explicitly and accompany them with the basis upon which they were made;
— demonstrate and detail a commitment to a well-defined programme for monitoring project effects.

For those countries which are signatories to the Convention on Biological Diversity, a clear commitment is required as described in Article 14 of the Convention:

> Each contracting party, as far as possible and as appropriate shall:
>
> (a) Introduce appropriate procedures requiring environmental impact assessment of its proposed projects that are likely to have significant adverse effects on biological diversity with a view to avoiding or minimising such effects and, where appropriate, allow for public participation in such procedures;
>
> (b) Introduce appropriate arrangements to ensure that the environmental consequences of its programme and policies that are likely to have significant adverse impacts on biological diversity are duly taken into account.

There have been many EIAs for road projects. However, the biological and ecological components of these assessments are usually very limited in scope and the information often based on superficial surveys (see the examples in Appendix III). Usually more attention is paid to landscaping than to ecological assessments and ecological restoration.

Ecological considerations have been included within EIAs of some roads; for example, Box and Forbes (1992) have suggested a conceptual framework for an ecological input into road projects. Perhaps not surprisingly, given the popularity of the concept of biological diversity since the 1992 Convention on Biological Diversity, there are some assessments of impacts on roads which focus in particular on biological diversity. For example, Southerland (1995) has proposed that an ecosystem approach is critical to assessing biological diversity effects and should include *inter alia* establishing biological diversity goals and end-points. Categories of end-points include 'consistency with regional plans', 'functional integrity of regional ecosystems' and native 'species diversity'. An example of a biological diversity conservation goal could be to maintain native diversity and natural processes. The end-points could include native species diversity, native structural habitat diversity and status of hydrology, nutrient and energy cycling, fire regime and keystone species interactions. Once back-ground information has been obtained, the effects of the road project on biological diversity can be assessed. Southerland (1995) indicates the need for careful evaluation of the effects and each alternative for attaining the biological diversity goals and objectives (Table AII.2).

AII.3 ECOLOGICAL IMPACT ASSESSMENTS

Ecological Impact Assessments are considered in some detail here because this book is about the biological and ecological effects of roads and traffic.

An Ecological Impact Assessment (EcIA) is a formalised process for identifying and quantifying impacts of a proposed project on ecosystems, their components, properties and processes. An EcIA may be used in conjunction with an EIA or be part of an EIA along with economic assessments and sociocultural assessments.

Research on the methods for EcIAs has often been less extensive than for EIAs in general. However, there are now some fairly detailed texts on EcIAs. For example, the Institute of Environmental Assessment (IEA, 1995) has published 'Guidelines for Baseline Ecological Assessment'. Schmitt and Osenberg (1996) edited a book containing many ideas about detection of ecological impacts in coastal habitats. Another book (Reijnen et al., 1995) describes methods for predicting the effects of motorway traffic on breeding bird populations. This is a particularly comprehensive volume from the Netherlands which includes very detailed theory and clear practical applications. More recently, J.R. Treweek's book *Ecological Impact Assessment* (1999) warrants attention.

Broadly speaking, an EcIA centres around some basic questions about the components, properties and processes of ecosystems:

— What is there in the area of the proposed road?
— Where is it (spatial distribution)?
— What is the abundance?
— What state is it in (its condition, integrity)?
— Are there elements of ecological or biological importance?

Identifying and then mapping land-use types and also habitats (as general categories or specifically on the basis of an ecological classification) is one way of building up a picture of what is there, its locality, abundance and

Table AII.2: Conserving biodiversity in highway development (from Southerland, 1995)

BIODIVERSITY END-POINTS	NO ACTION	ALTERNATIVES	
		ALTERNATIVE 1	ALTERNATIVE 2
Consistency with regional plans	Does not provide transportation level of service	Opens planned agricultural area to urbanisation	No conflicts with local growth plan or state highway plan
Functional integrity of regional ecosystem	No change	Loss of forested area in core natural area	No effect on core area
Composition of habitat types in region	No change	Reduces the proportion of old-growth forest in region	Does not significantly change proportion of natural habitats in region
Area of sensitive communities	No change	Loss of 100 acres of old-growth forest and 10 acres of wetland	Loss of 50 acres of second-growth forest along agricultural fields
Status of sensitive communities	No change	Open edge habitat created along 50 miles of old-growth forest	No change
Native species diversity	Existing exotic weed problem in second-growth forest	Possible invasion of exotic shrubs into forest edge	Includes exotic plant management
Native structural habitat diversity	No change	Loss of old-growth forest, removes snags	Loss of second-growth forest removes hedgerows
Status of hydrology, nutrient and energy cycling, fire regime, and keystone species	No change	Reduction of subsurface flow to adjacent wetlands	No significant change in hydrology
Number of sensitive species	No change	Local extinction of wetland species	No change
Status of sensitive species populations	No change	Nest parasites reduce reproductive success of forest interior birds	Loss of foraging area for some forest species
Habitat connectivity	No change	Severs 5 parcels of old-growth forest	Eliminates hedgerows connect-

Contd.

Table AII.2: (Contd.)

Habitat patch distribution	No change	Fewer large forest patches	ing second-growth forest No change in proportion of patch sizes
Number of contiguous habitat areas affected	No change	Disturbance radius (16 km) intersects 2 wilderness areas	No change

state. Broad categories such as pastoral land, protected areas, recreational land etc. are common. A general habitat category would be wetland or woodland. Both woodlands and wetlands can be further classified on the basis of selected biogeophysical criteria.

The overall structure of an EcIA can be based on a number of stages as follows (see IEA, 1995; Treweek, 1999 for details:

- The legislative context—what is legally required? Land ownership. What is legally required for an EcIA.
- Public consultation—who to consult and how and when?
- Scoping or early screening—defining the context, spatial and temporal parameters and limits to the EcIA.
- Focusing (deciding what can be done within the time and resources available).
- Identifying and quantifying impacts.
- Identifying mitigation and compensation.
- Ecological monitoring.

A particularly important stage is that of scoping or early screening. If well done, the scoping stage can provide a sound basis and save time and resources. Scoping includes a process by which the spatial boundaries for a road or at least agreement on them is reached. Additional questions might be: What area if any beyond the physical boundaries of the road should the EcIA consider? What associated road construction processes should be included in the EcIA?

Assessment of environmental effects during construction of roads have only occasionally been included in EIAs. Reports in the literature concerned only with effects during construction are few. For example, Burke and Sherburne (1982) monitored wildlife populations and activity during and after construction of a road in northern Maine. They reported that movements and densities of birds and mammals did not differ significantly during and after construction. Cramer and Hopkins (1982) examined the effects of a dredged highway construction on water quality in a Louisiana wetland. They reported that any effects tended to be temporary and that increased turbidity and colour gradually returned to preconstruction conditions.

There needs to be agreement about the time available and when the EcIA is to take place. The proposed road may have a direct impact on some species and habitats but over what area and during what times (year) will the EcIA be undertaken. There needs to be a balance between that time available and the depth of the EcIA in terms of scientific integrity.

It is not practical to undertake a comprehensive EcIA where every species is taken into consideration and where every ecosystem component, property and process is examined. Therefore, it follows that selection is necessary and criteria must be agreed to for that selection. Priority species would include those prior protected, species in decline and those categorised as rare or endangered.

Environmental monitoring and ecological monitoring should follow from EIAs or at least be included in EIAs. Very few reports deal with this aspect, especially ecological monitoring. Two papers (Pratt and Coler, 1976; Davis and George, 1987) describe monitoring of urban and road runoff.

AII.4 ESSENTIAL CONTENTS IN AN ECOLOGICAL IMPACT ASSESSMENT

The following should be addressed.

i) Statutory requirements

What legislation applies to road construction and the ecological effects of roads and traffic?

Is an EcIA legally required for a road project?

Identify the statutory requirements for an EcIA. Are there any international agreements? Is there a requirement to meet the needs of the Convention on Biological Diversity? Is the road project likely to impact on sites protected by international agreements?

Does the national legislation include protected areas and/or protected species? If so, are there protected areas and/or species in the area proposed to be taken up by the road?

What is considered 'significant' in terms of likely impacts on the environment?

ii) Consultation

Are there any organisations that must legally be consulted (such as planning authorities, conservation authorities)?

Identify the interested parties or stakeholders (the individuals or groups who can affect or who are affected by the actions, decisions, policies or goals of the proposed development).

Establish the consultation process as early as possible with all interested parties.

iii) Scoping

Scoping concerns designing the EcIA based on information about the proposed project or development. Within the scoping exercise, the following are considered:
— Language, communication and definitions
— Proposed timetable for the development
— Protected habitats and protected species
— Identifying impacts and their causes
— Identifying the nature and extent of the ecological impacts
— Identifying constraints (administrative, financial and technical)
— What can be achieved in the time available for the EcIA (what is desirable, what is practical).

iv) Definitions of technical terms

Multidisciplinary teams and groups of people are involved in EcIAs within a general framework of EIAs. Not surprisingly, language presents a challenge. Simple terms such as 'natural', 'landscape' and 'environment' will mean different things to different people.

Terms such as 'environment', 'ecosystem', 'biological community', 'biological diversity' and 'populations' should be clearly defined to avoid misunderstandings at a later date.

v) Information technology (IT)

Ecological impact assessments may be undertaken at different scales using many data sets. The use of IT, and Geographical Information Systems (GIS), will almost certainly be an important element in spatial analysis and analysis of the interactions between the road and the environment. Computer-aided design is another tool that may be used to analyse visual impacts on the environment.

vi) Spatial extent of the impact

There needs to be agreement on the spatial extent of the EcIA. Agreement is also imperative on whether or not associated activities are to be included in the impact assessment. For example, does the EcIA include the ground occupied while quarrying local gravel for road building and temporary quarters for the road-building crew.

vii) Recognition of the functions and values of nature

Nature has intrinsic values but it also provides sources, sinks and services used by humans. Mangrove swamps, for example, provide a habitat for a range of aquatic species and support mangrove biological communities. The ecosystem helps prevent flooding of inland areas and provides a source of food and fuel for humans. Swamps are used to absorb wastes (the swamp is both a source and an ecological sink). Wetlands in general contribute to water quality and quantity management. They support wildlife and recreational activities.

viii) Sources of ecological information

How will the baseline ecological information be obtained? It may have to be commissioned and collected via surveys and the establishment of new databases (primary information) or taken from existing databases (secondary information).

Decisions have to be reached as to who obtains the primary information, who has the technical skills and who pays?

Secondary information: The sources of secondary information need to be identified. These could be in the form of national or regional biological databases. Alternatively, the information might be obtainable from councils, government departments, or local natural history organisations.

ix) Gathering primary ecological information

Many methods are used by ecologists for collecting primary ecological information. All these methods have their limitations as well as biases towards certain groups of organisms. The limitations and biases should be considered when the ecological information is analysed and presented.

Ecological surveys can be in the form of direct observation, that is, simply walking the route in a random fashion and recording what is seen. There are two main limitations here. This method will detect only those organisms, habitats and communities present and visible to the observer at the time of his walk. Typically this would include many kinds of plants (grasses, herbs, shrubs and trees), mammals, birds and butterflies. At the same time other information is required: topography, geology, soil types, aquatic features and climate.

At any moment in time there will be organisms not visible to the observer and therefore not recorded. Secondly, some organisms which are present and visible may be overlooked because of the random nature of the survey method.

Methods and equipment are available which enable more comprehensive surveys, whether for organisms in the soil, aquatic organisms or those highup in tree canopies. Similarly, there are survey methods which enable collection of information in a more systematic and compressive manner. One well-known and widely used technique demonstrates this point, namely use of the pitfall trap in surveying ground-dwelling invertebrates such as beetles. Pitfall traps are simple plastic coffee cups sunk into the ground, into which beetles and other small organisms fall and become trapped. Left in the ground and checked periodically, they may provide information on the species present in the area. The trouble is, not all species of beetles will be trapped this way. Some may fall into the traps and crawl out again. All survey methods have their limitations.

x) Surveys and sampling effort

Not surprisingly, the amount of 'sampling effort' will influence survey results. For example, the number of bird species recorded during a one-hour random walk along the proposed route is likely to differ from a series of random walks undertaken for two hours, one day per week, for three weeks.

xi) Temporal and spatial variation

Similarly, one random walk during a midsummer afternoon to record birds is likely to yield different results than a series of walks during different times of the day and night throughout the year. Animals come and go and some plant species such as spring flowering bulbs may be present at only certain times of the year.

The distribution and movements of plants and animals over time and in different parts of the landscape will have implications for the comprehensiveness and accuracy of the information collected. Several techniques enable addressing the temporal and spatial problems that arise in ecological surveys.

xii) Species

A common unit in biological and ecological studies is the species. For this reason, most ecological surveys centre on species (rather than on populations, subspecies, families or higher units of taxonomic organisation). Baseline information will therefore almost certainly include species. However, it is impractical to have a definitive account of all species due to constraints of time and technical expertise requisite for comprehensive surveys of all plants, animals and other organisms on the ground, in the soil and in water bodies. To be comprehensive, surveys would have to be undertaken throughout the year to accommodate seasonal changes in the activities of some organisms. There is likely to be lack of the specialised information necessary for identifying certain groups of organisms.

Because of this difficulty, very often ecological surveys do not attempt to obtain information on all groups of organisms. Commonly, the groups chosen are those more readily surveyed, such as species of plants, birds, butterflies and large diurnal mammals. The issue is not so much which groups are selected, but the criteria for selecting or not selecting a group.

The criteria could be based on any of the following:

— legally protected species;
— species in red data books (Red Data Books are lists of species compiled on an aggregation of ecological and other criteria; these are the species in greatest need of conservation);
— species considered or shown to be 'rare';
— keystone species (species that are keys to an ecosystem—their—extinction has implications for the survival of many other species);

- species with specialist requirements;
- species considered culturally important;
- economically important species;
- easily surveyed species;
- indicator species (those species whose presence, absence or condition tells us something about the state of the habitat or the environment);
- umbrella species (those species which if conserved, will benefit the conservation of other species using the same habitat).

A rationale and criteria should be provided for the selection of species to be surveyed.

xiii) Habitats and communities

Loss of habitats and fragmentation of habitats are the most serious impacts arising from road projects.

It may be possible to obtain reasonably comprehensive information on habitats and/or biotic communities within the proposed route for the road. However, the terms 'habitat' and 'biological community' are often used in a general and subjective manner. A habitat, ecologically speaking is that place where a population of a certain species lives, thus a kingfisher habitat. Within that habitat there will be other species and there may be more than one biological community. The reason for identifying the habitat is to ensure that if representatives of a particular species are to be conserved or not disturbed, the habitat needs to be protected.

A biological community is made up of a collection of plant and animal species. Thus we generally speak of wetland communities, woodland communities or coastal marine communities. However, any one of these general communities could be one of many kinds. In some parts of the world, particularly in Europe, biological communities have been classified into different types based in part on the particular mix or assemblage of species present.

Woodlands and scrublands in Britain, for example, have been classified into 25 different kinds. Of those, four kinds of oak woodlands have been recognised (Rodwell et al., 1991). Thus there may be more than just one type of wetland or woodland or grassland.

xiv) Ecological processes

Ecological processes include:
- litter (leaf fall etc.) accumulation,
- decomposition,
- respiration,
- bioaccumulation,
- productivity,
- nutrient accumulation—eutrophication.

It would be impractical to undertake an assessment of the effects of a road on all ecological processes. Hence a selection has to be made and prior to that some criteria agreed to for selection of the processes to be assessed. Such a process could be the bioaccumulation of metals in nearby aquatic communities which may receive water runoff from the new road.

xv) Data reliability and confidence limits

The fact that all ecological survey methods have their limitations has been noted. Whether the ecological information is complete or only representative of an area must be stated. In other words, the confidence level of the information and its reliability must be described in statistical or other terms. An independent review of the ecological survey and information gathered might also be considered.

xvi) Methods for identifying impacts

Impacts can be direct, indirect, associated, cumulative or synergistic. What methods can be used to ensure that the identification of impacts is as comprehensive as possible?

As mentioned above, the methods used for identifing impacts in EIAs and EcIAs include checklists, flow diagrams and matrices.

A checklist is simply a list of actions and a list of environmental characteristics. Obviously, the more comprehensive the list, the more thorough the impact assessment. A checklist may be simply a list of environmental impacts caused by the road project or/and also a list of ecological components and processes.

Flow diagrams help to identify secondary and tertiary effects and interlinks among project impacts. For example, a sealed road increases the area of impervious surfaces, which in turn changes the rate and volume of water runoff. Water runoff may be channelled away from the road and directed perhaps to a river or estuary. Sudden increases or large volumes of fresh water entering an estuary may have an affect on some estuarine animals.

Matrices come in many forms and are basically a list of project actions opposed to a list of ecological components and processes. The matrix is an aid to identifying likely impacts.

Matrices, checklists or flow diagrams tell us nothing about the magnitude of the impact. They simply help to identify it. The magnitude of effects can be quantified in various ways but can also be assessed in a subjective manner and scored out of an agreed range of points. For example 0-10 could be the scale and 10 could be the greatest magnitude. The obvious disadvantage of subjective scores is their subjective nature. The advantage is that they help to deal with a wide range of information in limited time and facilitate communication of impacts to a wide audience.

Alternatively, impacts can be quantified. For example, the area of loss of a particular plant community as a result of road land-take can be measured. More difficult are measures of losses of numbers of plants and animals as a result of habitat loss. Even more difficult is the effect of physical disturbance on nearby wildlife when the road is completed and operational.

xvii) What is significant and how to decide priorities?

Are all effects equal in terms of their importance or are some relatively more important than others? Is the loss of a heritage site of greater importance than the loss of a habitat or a rare species. Is the loss of an ancient woodland greater than the loss of an ancient church. These are issues that need to be discussed and there has to be an agreed rationale and process for deciding priorities.

xviii) Ecological mitigation and compensation

Ecological impacts should be avoided whenever possible. If not, then proposals for remedying the effects or mitigation and compensation measures must be devised, presented, assessed and adopted.

Ecological and conservation considerations come first but have to be taken in a wider context of what is practicable and not just what is desirable. The monetary costs of such measures must be considered along with the environmental and conservation costs.

xix) Ecological monitoring

What needs to be monitored, how often and by whom?

Remedying and mitigating the effects may necessitate some follow-up studies. Monitoring ecological change over time is a commitment that requires not only ecological expertise but also an infrastructure. Ecological monitoring means setting objectives so that performance can be measured. A framework for establishing objectives is shown in Table AII.3.

xx) Communicating results

Who is the intended audience and how best to present the information?

An EcIA results in much information, some technically complex. Not all the interested parties in an EcIA may have the expertise to readily understand all the information. Consideration must therefore be given as to how the information is to be communicated be that written or verbal. A table of data may be useful to those with technical expertise. Converting such tables to other forms or aggregating data into a single index are just two of the methods that might be employed.

Table. AII.3: Theoretical framework for establishing objectives in ecological monitoring programmes (from Spellerberg and Sawyer, 1996).

Spatial Scale		Natural	What is possible	Legal	Significant Global/Regional/Local	Vulnerable Global/Regional/Local
Landscape level	Composition	Native flora and fauna with wide distribution, % broadleaf	Examine land base for topographic influences, e.g. watershed boundaries	Geological SSSIs 5% Broadleaf policy	Archaeological sites of importance, e.g. barrows	Habitats, e.g. heathland, mires etc.
	Structure	% open space % long rotation	Dendritic pattern of 1st, 2nd and 3rd order streams/rivers	Council structure plans.	Landscape views of importance	
	Function	Maintain landscape connectivity Natural large-scale disturbance regimes	Dependent upon the above	Maintaining migration routes for anadromous fish	Maintenance of clean, healthy water supply to downstream users	Ecological flows across ecotones and landscape boundaries
Stand level	Composition	Native flora and fauna with local distribution	Based upon BEC edaphic grid (soil nutrient and moisture content)	Species protected by W&C (1981) Act	Species in CITES, Appendices I and II Species of value in the local market	Organisms requiring forest interior conditions or continuity of habitat conditions, e.g. old growth
	Structure	Dead wood —snags and	Constraints on thinning and		Trees with Tree Preser-	Deadwood removed for use as firewood

Contd.

Table. AII.3: (Contd.)

Spatial Scale	Natural	What is possible	Legal	Significant Global/Regional/Local	Vulnerable Global/Regional/Local
	CWD	planting and other silvicultural regimes because of windthrow hazard etc.		vation Orders Structures which are valuable habitat to rare or endangered organisms, e.g. bat roosts	
Function	Small-scale disturbances	Maintain soil productivity			Nitrogen fixing

Appendix III

Examples of Environmental Impact Assessments

INTRODUCTION

In this section there are three examples of environmental impact assessments which show the range of details included in such assessments. They range from minimal informa-tion to well-documented ecological effects with suggestions for avoiding or compensating those impacts. These examples are direct extracts (references cited are not included in the list of references of this book).

Example one is from the Environmental Impact Assessment for the upgrading and realignment of Bruce Road, Tongariro National Park, New Zealand, prepared and published by the Department of Lands and Survey, Wellington. It is taken from a 1985 publication and is a typical example of the very simple and limited biological information included in many EIAs of that decade.

Example two is the SH2 White Pine Bush realignment: landscape and ecology assessment of effects report. Transit New Zealand, Hamilton. Reproduced with permission of Opus International. It is taken from a 1997 publication. The biological information is fairly extensive and the ecological information usefully includes 'potential adverse ecological effects'. A general direction is given with respect to ecological mitigation.

Example three is from the Final Environmental Impact Statement for the Beaver Basin Rim Road, Pictured Rocks National Lakeshore, Michigan. Reproduced with permission of the U.S. Department of the Interior. It is a 1996 publication and contains levels of detail which should be considered minimal for the alternatives, the affected environment and environmental consequences.

EXAMPLE 1

Vegetation

The sites directly affected by the Bruce Road upgrading proposal include representative areas of the normal subalpine to lowalpine vegetation mosaic on the north-western slopes of Mt. Ruapehu. The vegetation associations (between 1250 m and 1500 m) are characterised by absence/presence of ash soil cover, water table and drainage condition, shelter and aspect, and normal altitudinal distribution limits.

The effects of the road realignment on the vegetation can be summarised as:

(a) No unique or significant vegetation type or plant is threatened by the immediate effects of the roading proposals. An area of low-alpine red tussock-Mountain inaka will be lost below the '2nd Bluff'. This vegetation type is widespread elsewhere but the small area concerned is unusual in the immediate vicinity.

(b) Indirect effects of road maintenance and construction already threaten a bog area below the '1st Bluff'.

(c) The ground cover of the affected plant communities is determined by completeness of ash cover and its water-table characteristics.

(d) Colonisation of bare ash soil and loose scoria is naturally inhibited by frost action. Large boulders offer protection and soil traps for colonising plants. Lava outcrops and roading gravels lacking soils are not readily colonised.

(e) Natural erosion of vegetation on lava outcrops is largely confined to small areas subject to surface runoff, and bare scoria and sand subject to frost heave.

(f) On ash soils, subalpine vegetation is inherently stable on all slopes but is unable to establish where surface runoff buries or bares the ground surface as most recolonisation is by seedlings. Higher, in the lowalpine zone, plants often recolonise by prostrate growth or asexual spread, so soils will eventually be revegetated if protected from frost action.

Subalpine ground stripped of ash is therefore recolonised by a different plant community and is unlikely to return to the surrounding vegetation type.

(g) An area of moss below the 2nd Bluff will be covered with fill when the radius of corners is increased (Figure, after given).

(h) Some loss of tussock grassland habitats and mountain tarns will occur in the lower sections of the proposals along the 1973 deviation option.

Fauna

None of the proposals are expected to have any impact on the existing fauna identified within the Project Area.

Landscape Impacts

Landscape character

The project area is basically one of physical change and seasonal extremes and consequently any alterations to the landscape accelerate this natural process. Landform and surface cover identify the volcanic and alpine nature of the area and must be protected during construction work. The proposed upgrading is through an area which has already been significantly altered to accommodate the existing road alignment and further impact on the overall character of the area will be minimal, with the opportunity to ameliorate previous damage. If Stage III, Option II is adopted, care should be taken to preserve the integrity of the undulating landforms and flat alpine herb fields by following contours and working within a strict construction corridor. Guidelines for construction will be developed around the principle of minimal impact to the vegetation and surface materials.

Visual qualities

The unique visual qualities of the Project Area relate to the volcanic nature and appearance of the landforms and surface cover. Any cut-and-fill will create artificial forms unnatural to the area and will need to be modified so that they blend as much as possible into existing contours.

Major cuts in the 1st Bluff (Stage III, Option I) area will have considerable visual and physical impact because of the geological structure of this area. The 1st Bluff road alignment is cut through a lahar base area of loose rock and ash, which is prone to erosion. Surface water erodes the top of the embankments undercutting them.

EXAMPLE 2

Description of Ecological Features

Scientific names and conservation status for flora and fauna species found or likely to be found in the vicinity of the proposed realignment are described in Appendices II and III.

Flora

Two remanent indigenous plant communities dominate the study area— remanent broadleaf/podocarp forest within White Pine Bush Scenic Reserve and regenerating forest/shrubland on the fringes of the eastern side of the highway within the Regional Council—administered Soil Conservation Reserve.

GENERAL VEGETATION DESCRIPTION

Most of White Pine Bush Reserve is covered in native forest. Its canopy is mainly broadleaved species which have regenerated following logging within the last century. The canopy is dominated by tawa, titoki, rewarewa and lacebark, with some hinau. Mature emergent kahikatea and a few matai and rimu are scattered throughout the reserve. An exceptionally diverse epiphyte community covers the upper reaches of these large podocarps.

Young kahikatea are common in the canopy in the north of the reserve. On the reserve's other margins kohuhu, mahoe, putaputaweta, tree tutu, cabbage tree and pate are predominant. Beneath the canopy is a dense understorey of broadleaved species (mahoe, kawakawa, pate, hangehange, pigeonwood and tree fuchsia), tree ferns (mamaku and wheki) and occasional ongaonga and young nikau. The dense ground cover consists mainly of ferns such as kiokio, hen-and-chickens fern and *Pneumatopteris penniger*. The forest floor has prolific generating seedlings of the canopy tree species, particularly titoki. Supplejack entanglements occur throughout the forest from the ground to the canopy.

Dense bracken fernland occurs on the western boundary. It contains areas of rank pasture and trees and shrubs of kanuka, manuka, mahoe, karamu, lacebark, titoki, kohuhu, cabbage tree and ngaio.

Rank pasture occurs at the north-western and south-eastern edges. It consists mainly of exotic grasses and herbs.

At the south end of the reserve is a picnic amenities area. Mown grassland is surrounded by planted exotic trees (mainly *Robinia*, elms, willows, poplars and eucalypts) with some planted native trees (lacebark, ngaio, lemonwood, ribbonwood, kahikatea and *Pittosporum ralphii*).

Grazed pasture occurs within the reserve on its north-western margin, and stock-proof fences lie inside the reserve boundaries.

Vegetation Within the Proposed Realignment—White Pine Bush

Within the proposed alignment footprint, healthy second growth titoki forest is regenerating to titoki dominant lowland forest. There are signs of some goat browsing and possum presence. The canopy over the majority of this area ranges in height from 8-14 metres and includes trees up to about 30 cm at breast height (dbh). The canopy is closed over most of the area, the exceptions being a few small areas of grass and blackberry adjacent to the road.

Three large kahikatea ranging in diameter from 0.75 m to about 1.5 m dbh and about 25-30 m tall emerge from this canopy. Several smaller kahikatea trees (10 to 15 m tall and dbh up to 30 cm) are also present directly adjacent to the current highway on the upper north-east side of the reserve.

The canopy contains a fairly even mix of tree species which includes titoki, lemonwood, mahoe, cabbage tree, lacebark, ngaio, kohuhu, pigeonwood, marble leaf and mamaku. Koromiko, tutu, flax and pampas are growing closer to the roadsides.

As the distance from the existing road increases so do the canopy height and tree diameters. Towards the Kareaara Stream, the second growth forest becomes tall mature titoki/podocarp forest. Several large (30 m plus) kahikatea and a large rimu dominate. A subcanopy of small titoki and rangiora is present in places. The understorey consists of shrubs such as kawakawa, hangehange, mahoe, rangiora and pigeonwood on the drier sites and a mix of these shrubs and ferns on wetter sites and on the stream banks. Dominant fern species are tree ferns, hen-and-chickens fern, *Blechnum capense* and *Pneumatopteris pennigera*. There is a moderate to dense ground cover of hookgrass, tree seedlings and leaf litter. In places there are high numbers of titoki seedlings and a few kahikatea seedlings.

Soil Conservation Reserve

On the eastern side of the existing road a mixture of common regenerating indigenous and exotic plants dominate the ungrazed and non-production forestry areas. No ecologically significant indigenous plant communities or species are present. Self-seeded *Pinus radiata* are present.

Fauna

Birds

Tui, bellbird and kereru appear to utilise the study area on a seasonal basis. Other indigenous birds inhabiting the study area include fantail, grey warbler, silvereye and kingfisher.

Although there have been recordings of North Island brown kiwi in White Pine Bush, the last confirmed bird left in the bush has not been heard or sighted since the mid-eighties. There are now no kiwis left in the locality (J McLennan, *pers. comm.*).

No other threatened or rare bird species are known to utilise the forest remnants.

AQUATIC BIOTA

A small (1 m wide and 0.2-0.5 m deep) unnamed stream runs through the forest/shrubland before entering the Kareaara Stream, which the existing highway crosses, within White Pine Bush. The lower reaches of this stream have a soft silty substrate, covered by large limestone boulders, characteristic of small streams in the area. The upper reaches contain a more rocky/gravel substrate. The entire stream is enclosed with a dense canopy of indigenous vegetation.

A brief inspection of the stream indicated that the upper catchment provides habitat for a variety of aquatic macroinvertebrates such as mayflies, stoneflies, caddis flies and chironomids. Freshwater snails are also likely to be present in the lower reaches. The presence of net-spinning caddises and stonefly nymphs suggest a healthy aquatic community.

Although no site specific fish survey was undertaken, stream morphology and habitat characteristics suggest both short-finned and long-finned eels, banded kokopu, koaro and bully species may be present (McDowall, 1990). However, the small size of the stream combined with the fish barrier created by the Tangoio Falls will limit exotic trout and galaxiid habitat utilization.

Significance Assessment

Landscape

In relation to the report's Northern Hills Landscape Unit, White Pine Bush is not seen as an Outstanding Landscape Feature nor as a Special Landscape Character Area. From the combined sum of the landscape assessment matrix, the area has a rating of 16 in a range of 10-25 for the landscape unit.
Significant Landscape Areas and Features that are defined as outstanding (Lake Tutira, Tutira Basin and Maungahururu Range, Titiokura Saddle—Te Waka Range) all have ratings of 24-25. This implies that while White Pine Bush is a significant or definable landscape, it does not have components which elevate it to being outstanding.

Ecology

WHITE PINE BUSH

White Pine Bush is one of the last remnants of lowland forest left in Hawke's Bay. The mosaic of mature kahikatea forest combined with mature titoki forest is exceptionally rare within North Island. Less than 2% of this type of forest remains within the lowland area of North Island today. It also contains a 'locally threatened' epiphytic orchid—*Bulphylum tuberculatum* (Cameron *et al.*, 1995).

The forest remnant has moderate wildlife values but is important from a district perspective. It also provides habitat opportunities for freshwater fish, including important galaxiid fish species. Finally the forest may have important wildlife corridor functions for other fauna within the locality (Ogle in Saunders *et al.*, 1987).

Therefore White Pine Bush is considered to be of regional, if not national significance and has a 'Sites of Special Biological Interest' ranking of 'High' (Kessels, *pers. obs.*).

Soil Conservation Reserve

The forest/shrubland remnants, while relatively modified, are regenerating well. The remnants also afford some wildlife corridor and riparian marginal values for the locality. Consequently the forest/shrubland area is considered to have a Sites of Special Biological Interest ranking of 'Potential' and is of district/local significance (Kessels, *pers. obs.*).

Assessment of Landscape and Ecological Effects

Landscape effects of the proposal

Vegetation removal and earthworks associated with the proposal will have an obvious visual effect, particularly where the resultant batters are visible from a distance and from the adjoining reserves. The proposal will involve initial removal of vegetation from the sections of realignment and then excavation of two large bulk cuts along with two smaller sideling cuts and placement of a large fill.

The proposed bulk cut south of Morse's Curve on the east side of the existing highway will be approximately 45 m high by 220 m long at the base. This will necessitate removal of vegetation within the Soil Conservation Reserve from the existing road edge up to the pine plantation above. The Tangoio Walkway will also have to be realigned upslope. Relative to the final landform and highway alignment, this batter will be visible to north-bound traffic from Karaaera Stream Bridge, approximately 400 m to the south and to south-bound traffic from the proposed Tangoio Headquarters Curve, approximately 700 m to the north. A flat open area between the existing highway and the proposed alignment will also be created.

The proposed bulk cut at Tangoio Headquarters Curve on the east side of the existing highway will be approximately 12 m high by 130 m at the base. As with the previous cut, this will require vegetation removal within the adjoining reserve and the possible realignment of the walkway. This cut will be visible to north-bound traffic from the Morse property entrance, approximately 150 m to the south and to south-bound traffic from the entrance to the White Pine Bush Reserve car park. The cut will also be visible from the car park itself and the adjoining picnic shelter.

The proposed White Pine Bush Curve realignment will entail a sideling cut and a bulk cut on the east side of the existing highway, the southern sideling cut being 10 m high by 70 m at the base and the northern 20 m high by 100 m at the base. The fill will be placed on the west side of the existing highway between the two cuts. The fill will be 40 m high (slope face in the order of 35 m high), a maximum of 60 m wide at the base at its midpoint and 200 m long at the centre line. This will require removal of vegetation, placement of a new culvert and a section of retaining wall within the fill batter to protect two mature kahikatea trees and the adjoining walking track and associated bridge—all within the White Pine Bush Scenic Reserve.

The southern sideling cut will be visible to north-bound traffic from the entrance to the Scenic Reserve car park, approximately 100 m to the south and to south-bound traffic from an existing curve, approximately 700 m to the north. The northern sideling cut will be visible to north-bound traffic from the proposed Tangoio Headquarters Curve, approximately 800 m to the south and from the existing curve, approximately 200 m to the north. The western side of the fill will be visible to traffic on the immediate approaches to the curve and from the walking track within the reserve.

From within the reserve it is likely that parts of the fill batter will be visible from the walking track as it approaches the local stream. In the area of the walking track bridge the toe of the fill batter will be approximately 19 m from the track. The two mature kahikatea trees will remain 5 m upstream of the bridge but closer to the fill batter and section of retaining wall there will be no vegetation in the short term due to construction. Elsewhere in the immediate area vegetation will remain between the track and the proposed alignment; the depth of this vegetation will increase as the fill tapers away from the track.

SUMMARY OF LANDSCAPE EFFECTS

Within the enclosed local landscape the vegetation removal and earthworks implied by the proposal will be obvious to traffic travelling between the Kareaara Stream Bridge and an existing curve to the immediate north of the proposed fill batter, a distance of approximately 2.5 km. There will be no visible effects on traffic beyond this length.

The landscape effect on White Pine Bush Scenic Reserve and its users will result from land severance, removal of vegetation and visibility of some of the earthworks.

The landscape effect on the Soil Conservation Reserve and its margins will result from land severance, removal of vegetation, realignment of sections of the Tongoio Walkway and visibility of some of the earthworks.

All of the landscape effects will be obvious for the short- to medium-term future. The proposed bulk cut south of Morse's Curve will be the most obvious effect due to its scale and proximity above the highway.

Potential adverse ecological effects

Potentially significant impacts of the proposed road realignment on the forest remnant, stream and forest/shrubland areas are likely to be:

Direct impacts:

— habitat loss and damage, and destruction of plants and wildlife habitat, in the course of road construction;
— degradation of water quality and hence aquatic biota habitat quality.

Indirect impacts:
- loss, damage, modification or fragmentation of adjacent forest remnants;
- disturbance either from highway traffic, reduced breeding success of individual birds nesting in close proximity to the road;
- increased colonisation of indigenous plant communities by adventive weed species such as wattle on exposed batters and cuts of the new road.

Indigenous vegetation removal

WHITE PINE BUSH

The approximate area of indigenous forest destroyed by the proposed realignment is 0.5 ha (of the total 19 ha area of reserve). This equates to about a 2.6% loss.

The vegetation lost due to the proposed road and associated embankments is a mix of mature titoki forest and regenerating second-growth forest with 3 large emergent kahikatea trees located about 10-15 metres from the road. The design of the road was such that direct impact on these kahikatea was avoided. Further, the footprint of the fill batters has been steepened as far as practicable to minimise the amount of forest lost and to reduce impact on the root zones of several mature trees near the current reserve walkway and also at the north end of the reserve, where several smaller kahikatea are present.

Nonetheless, the proposal will still result in a substantial loss of forest. Unless comprehensive avoidance, remediation and mitigation measures are adopted, this loss represents a significant impact on a regional scale, bearing in mind the ecological value of the remnant.

REGENERATING FOREST/SHRUBLAND AREAS

On the other side of the road, a mixture of common indigenous and regenerating indigenous and exotic plants characterise the area within the proposed cuts. A small amount of this vegetation (about 2000 m^2 in total) will be removed by the proposed realignment and road widening. The ecological effects of this activity are considered minor.

Indirect effects on indigenous vegetation

While the direct effects of vegetation removal are easily quantifiable, the indirect effects are more difficult to assess. The three major indirect potential effects are associated with damage to root mats of the remaining kahikatea, titoki and rimu trees, changes in existing microclimates and intrusion of exotic weed species that exposed road batters often accelerate.

ROOT DAMAGE

Damage of the root systems of the mature kahikatea trees should be minimal if the fill batter edge is no closer than 10 m from the base of the trees (P. Champion, *pers. comm.*). Use of retaining walls in critical areas will further ensure reduction in risk of root damage to large trees. Sapling kahikatea trees roots may be damaged near the road batter at the northern end. However, the feeding roots around these podocarp species are close to the base of the tree. The trees are well established and thus unlikely to suffer significant impacts apart from a possible short-term reduction in growth rate.

MICROCLIMATIC EFFECTS

Adverse effects associated with edge effects, such as increases in adventive plant establishment and changes in microclimates could be detrimental if suitable mitigation measures are not undertaken (refer to section 7 for suggested mitigation measures) (Given, 1994).

It is generally acknowledged that seedling regeneration of kahikatea requires periodic flooding of silt-laden water and high water tables (e.g. Champion, 1988; Timmins, 1988). Flooding deposits silt, which increases soil fertility and induces development of secondary indigenous vegetation which favours the establishment of podocarps (Beveridge, 1980). However, as the prolific growth of kahikatea on this site indicates, high water levels resulting from seasonal flooding is only one of a number of factors influencing kahikatea regeneration. Other key factors influencing the establishment of kahikatea seedlings are light intensity, soil moisture content, presence of a tree root-mat and chemical inhibition of young kahikatea by mature kahikatea (Champion, 1988). Birds also play an important role through seed dispersal (Beveridge, 1980).

The proposed realignment may in fact enhance the growth and regeneration of kahikatea as backup of water behind the new fill batters in flood events will result in wetter conditions upstream.

POST-CONSTRUCTION INVASION OF EXOTIC WEEDS

The greatest potential threat to the integrity of the remaining forest is invasion of exotic plants such as old man's beard. This species, as well as a host of other weeds, are already present. Fresh earth exposed during construction will provide ideal conditions for the spread of these weeds. Therefore it is critical that an intensive post-construction revegetation and long-term maintenance programme be developed for this proposal.

Aquatic biota

During construction a large area of exposed cut-and-fill faces as well as direct work within the Kareaara Stream and its tributaries will create significant potential for suspended sediment discharges. Ryan (1991) and Quinn *et al.* (1992) suggest that adverse effects on aquatic biota associated with sediment discharges can be significant in the long term.

While the tributary may provide few habitat opportunities for kokopu, the Kareaara Stream does. Thus sediment runoff downstream of the works could affect this fishery. However, provided sediment runoff during construction is controlled using adequate silt control measures, effects are expected to be minor and short term.

It is considered that culvert design need not cater for fish or invertebrate passage, as the small tributary affords little fish habitat opportunity in its relatively small upper catchment.

Fauna

From a regional perspective the loss of indigenous bird habitat is not significant. However, from a local perspective, the loss appears potentially significant. Therefore some effort towards habitat replacement needs to be considered. As the existing highway already affects potential wildlife corridor features, the new alignment is unlikely to worsen the situation provided the old road is replanted with suitable indigenous plants.

Modelling suggests noise associated with the new alignment may be greater than the current noise levels. However, personal experience suggests that most common forest indigenous bird species roosting or nesting close to roads quickly adapt to noise disturbance.

Conclusion and Mitigation Recommendations

Conclusion

The proposed realignment of State Highway Two through the White Pine Bush area represents a significant potential adverse landscape effect at a district level and a significant potential adverse ecological effect at a regional level.

Ideally the proposed route should avoid any intrusion into the White Pine Bush Scenic Reserve. However, if social and economic advantages of the proposal are seen to outweigh the ecological disbenefits, then substantial remediation and mitigation measures are required to ensure the adverse effects will be minor.

Avoidance, remediation and mitigation recommendations

DESIGN PHASE

Avoid as far as practicable intrusion into White Pine Bush when setting the alignment.

- Minimise the area of indigenous forest cleared (i.e., keep roadside batters and cuts as steep as possible without precluding revegetation and slope stability).
- Design culvert installations to minimise impact on the natural streambed.
- Develop a detailed sediment control plan, recreation/landscape rehabilitation plan and environmental management strategy for the proposal.

These documents should be written into the construction contract documents. Non-compliance with these documents should result in financial penalties being imposed on the contractor.
- As part of the landscape plan design new vehicle and walkway access points and car parks in consultation with DOC. This may also include enhancement of existing recreational facilities.

CONSTRUCTION PHASE

- Avoid machinery contact with the reserve outside of the construction zone and stream-bed as far as practicable.
- Restrict clearance of native vegetation within the designation to the minimum required for the carriageway.
- Utilise the best practicable construction methods to minimise generation of silt and sediment. For example, geotextile material could be placed along the toe of all batters and cuts, temporary hay bale swales could be installed at all discharge points, stagger works so that only short lengths of fills or cuts are exposed at any one time, place erosion control blankets on critical fill faces near waterways, use rock rubble directly below the culvert outlet to act as a water energy dissipater.
- The cut batters should be hydroseeded with a hardy grass seed mix immediately after final shaping to preclude surface erosion and to initiate their rehabilitation. The fill batters and the flat area at the south of Morse's Curve should be topsoiled and oversown with grass for the same purposes. The seeding process will be repeated as necessary at a later date to ensure grass establishment and 'security' of the slopes.
- Indigenous trees and shrubs, grown from seed sourced in the area, should be planted into the initial grass cover of the fill batters and any suitable flat areas.
- Mitigation for loss of indigenous forest needs to involve rehabilitation of the unused sections of the existing highway and planting of exposed fill batters. The size and extent of this restoration needs to reflect the amount of indigenous vegetation removed.

OPERATIONAL PHASE

- Maintain any restoration planting programmes.
- Develop an environmental monitoring programme to monitor post-construction effects.

The details and extent of specific landscape and ecological mitigation, restoration and monitoring programmes should be developed in consultation with the Department of Conservation and Hawke's Bay Regional Council.

EXAMPLE 3

Vegetation

- The National Park Service, in consultation with the state of Michigan, the Soil Conservation Service, Hiawatha National Forest, and academic institutions with reclamation experience in the region, would develop a native plant material revegetation plan prior to construction. This lead time would be needed to prepare plant materials if needed for reclamation. Examples of native plants that might be used to revegetate disturbed areas include the following: hairgrass, little bluestem, poverty oat grass, bush honeysuckle, bearberry, and big-leaf aster. Plant stocks would be propagated from seed sources originating in the park. Exotic plant control measures that should take place before, during, and after construction would also be included as part of the revegetation plan.
- Borrow and aggregate material sources would be inspected prior to construction to determine needed weed control measures.
- To prevent the introduction of exotic plants in the park, straw and hay bales would not be used for erosion control or seeding mulch, pending state approval.
- Revegetation and vegetation monitoring would continue for at least two seasons following construction.
- The road alignment would be sited to avoid pocket wetlands identified in the U.S. Fish and Wildlife Service's natural wetlands inventory map, taking into account implications of other construction impacts should the alignment need to be shifted.
- Vista clearing would only be performed at the two proposed overlooks. Selective thinning of understorey and removal of lower limbs on tall trees would be performed.

Wolves

- Based on studies, currently there are no known dens or rendezvous sites in the project area. Prior to construction of the proposed road, the National Park Service, in compliance with section 7 of the Endangered Species Act, would consult with the U.S. Fish and Wildlife Service (USFWS), and work with the Michigan Department of Natural Resources, Hiawatha National Forest, and other landowners to determine if any dens or rendezvous sites have become established in the project area, and to take measures necessary to mitigate potential impacts.

Floodplains and Wetlands

Neither of the proposed road corridors would cross permanent rivers or streams between Legion Lake and Twelvemile Beach. Thus, there are no areas where flooding would be a concern.

The U.S. Fish and Wildlife Service inventory of national wetlands has mapped Pictured Rocks National Lakeshore and adjacent lands. There are many small lakes, bogs, and marshes within the park boundaries, classified by the National Wetlands Inventory as primarily palustrine/open water, palustrine/forested, palustrine/emergent, and lacustrine/limnetic/open water. Most of these wetlands are in the Beaver Basin lowland area.

A few small wetlands are in or near the road corridors. Several bog pockets exist around Legion Lake, and a classical bog lake lies east of the Twelvemile Beach campground. H-58 passes through or borders several small palustrine wetlands by Kingston Lake. However, in the Legion Lake area, south of the road, several larger wetlands are present. The only wetlands within the shoreline zone and IBZ corridors are three small "kettle wetlands" located north and west of Kingston Lake. The road would be routed to avoid all of these wetlands.

Vegetation

Pictured Rocks National Lakeshore lies within the northern hardwoods/hemlock/white pine region of the eastern deciduous forest of North America. Elements of the boreal coniferous forest and central grassland formations are also found here. Over 560 species of plants have been documented in the park, of which 88% may be classified as native (Read 1975, cited in Loope 1992). A 1987 survey of lichens in the park identified 235 taxa and concluded that the park had a diverse lichen flora (Wetmore 1988). A more recent survey of bryophytes in and near the shoreline zone corridor identified 128 moss and 13 liverwort species (Glime et al. 1991). This corridor contains only a small percentage of the likely bryophyte taxa in the park because it is mostly forested and lacks several habitats that are usually rich in species.

In the mid-1800s, the most extensive forest type in the park area was a mixed deciduous/coniferous forest composed primarily of sugar maple, beech, yellow birch, and hemlock. In the eastern part of the park, east and north of Kingston Lake (Kingston Plains), white pine/hemlock was dominant. Indeed, the Kingston Plains originally supported one of the most extensive stands of white pine on the peninsula. Ring counts on some of the larger stumps indicate that the stand was 300 to 400 years old when logging began in the area (NPS 1980).

Logging, repeated fires and land clearing for agricultural purposes, and introduction of exotics e.g., Dutch elm disease, have significantly changed the forests of the park area. Cutting of pine began in the mid-1880s and continued into the early 1900s. Several slash fires subsequently burned over pineland areas. The open "stump prairie" of the Kingston Plains is believed to be a result of these events (NPS 1980, Loope 1992).

Most of the original mixed deciduous/coniferous forest in the park and adjacent areas has either been clear-cut or selectively cut. The most recent logging in the shoreline zone was by The Cleveland-Cliffs Iron Company (now

Shelter Bay Forests, Inc.) in 1963. Logging operations are still conducted by Shelter Bay Forests and the state forestry division in the inland buffer zone and the area that H-58 passes through outside the park. The greatest change in the mixed conifer/deciduous forest has been a decrease in average tree size: most trees today are pole to small timber size (NPS 1980, Loope 1992). Several species are less prevalent than in the prelogging era, including yellow birch, sugar maple, and hemlock.

The Vegetation map shows the present general vegetative cover in the alternative corridor areas, based on LANDSAT imagery.[1] About 80% of the park is dominated by upland second-growth northern hardwoods. About 10% of the park is dominated by red, white, and jack pine (pine barrens or pine woodlands) on coarse outwash and coastal sands. Large successional stands of paper birch and aspen are also present on these soils. A relatively large white birch forest is located near the Twelvemile Beach campground. The remaining 10% or so of the park is dominated by boreal forest in wetter areas.

Within the alternative road corridors, no remnants of the original forest exist. The predominant forest cover types are mixed northern hardwoods/coniferous, pine woodlands, and pine barrens. The shoreline zone corridor is primarily in the mixed northern hardwoods/coniferous (79%) and pine woodlands (21%) cover types. The Fox River Pathway, about 1 mile southeast of the mouth of Sevenmile Creek, forms a fairly distinct boundary between the two cover types in the shoreline zone corridor. To the west of this line, the vegetation is generally dominated by hardwoods with some hemlock, while to the east it tends to be coniferous with patches of hardwoods present.

Dominant species in the mixed northern hardwoods/coniferous cover type are American beech and sugar maple, with some red maple, yellow birch, hemlock, and white pine. Some representative ground layer species include baneberry, spring beauty, wild sarsaparilla, jack-in-the-pulpit, yellow trout lily, sweet cicely, common wood sorrel, Canada yew, and club moss.

From a biological perspective, the pine woodland cover type is identical to the pine barren cover type—the same forest should grow in both areas. However, logging and fires have made the two areas much different today, and consequently they are differentiated on the Vegetation map. In the pine woodland cover type, in the eastern end of the shoreline zone corridor, a continuous forest canopy, with some isolated openings, is largely present. Red, jack, and white pine tend to be dominant species. Typical understory species of the pine woodland include blueberry, bracken fern, and huckleberry.

As shown on the Vegetation map, the IBZ corridor is primarily in the mixed northern hardwoods/coniferous cover type described above: 74% of the proposed road would be within this cover type. When the proposed road in the IBZ corridor stops paralleling the proposed road in the shoreline zone corridor and heads northeast across the Kingston Plains it would enter the

---

1. It should be noted that the cover types shown on the map are best estimates based on limited ground checking of the remote sensing information.

pine barrens cover type. About 26% of the total road alignment would be within this cover type.

Red, white, and jack pine are dominant trees found in the open pine barrens of the Kingston Plains. Unlike the pine woodlands, however, the trees are scattered in isolated clumps, or are in small tree plantations, in an open grassland or "stump prairie." Common hair grass, poverty oat grass, and reindeer lichen are the dominant species in the open areas. Other plant species include several lichen species, bracken fern, blueberry, and isolated clumps of pin cherry, hawthorne, quaking aspen, and white birch.

County road H-58 passes through all three cover types between Legion Lake and Twelvemile Beach. From Legion Lake to just west of the Adams Trail road intersection, H-58 is primarily in the mixed northern hardwoods/ coniferous cover type. When the road heads north across the Kingston Plains, it is largely in the pine barrens cover type. Near Twelvemile Beach the road is in the pine woodland cover type. Altogether, 61% of H-58 is within the mixed northern hardwoods/coniferous forest, 32% is within pine barrens, and 7% is within pine woodlands.

Several exotic plant species are present in the park along the road corridors. Spotted knapweed is found throughout the area. Other exotic species in the area include St. John's wort, purple loosestrife, burdock, and periwinkle.

Wildlife

Pictured Rocks National Lakeshore is located at the edge of the range of many eastern and western species and subspecies, and therefore the area supports a rich diversity of wildlife. Over 190 animal species have been identified or signs of their presence have been recorded in the park and adjacent areas. Between 1990 and 1992 four studies were conducted to survey invertebrates, amphibians and reptiles, birds, and mammals in the vicinity of the shoreline and IBZ corridors (with the exception of Kingston Plains) within the road corridor in the park. The results are summarized below.[2] No information is available on wildlife present along H-58, but this area would be expected to support similar species as in the park (with the possible exception of Kingston Plains).

Invertebrates

A total of 34 sites were surveyed for gastropods (a broad class of mollusks including snails, limpets, and slugs), on a 13-mile-long, 0.5-mile-wide corridor along the Beaver Basin Rim, during dry weather on July 23-25, 1992, and during rainy weather on September 26-27, 1992 (Pearce, *el at.* 1992). A survey

---

2. The results summarized here focus on the uplands sounth of the Beaver Basin rim, between Legion Lake and the Twelvemile Beach campground. It should be noted, however, that all of these studies also surveyed areas outside the road corridors, such as the shoreline from the Twelvemile Beach campground to the Hurricane River mouth.

for potential federally and state listed insects was also con-ducted on July 23-24, 1992.

Pictured Rocks National Lakeshore generally does not appear to support a large diversity or abundance of gastropods. A total of 33 terrestrial gastropod species were identified in or near the shoreline zone corridor. The cedar swamps adjacent to the shoreline zone corridor, along the base of the rim, appeared to provide the best habitat for terrestrial snails, with the greatest number of species in the greatest abundance. Pine-dominated and grassy plains habitats appeared to be poor habitats with fewer species and a lower abundance of snails. The beech-maple forest localities south and southwest of Sevenmile Lake were found to support more species and higher number of gastropods than localities to the north and northeast. The existing Beaver Basin overlook was found to support the most species (18) and had the fourth most abundant number of individuals recorded. The most numerous species recorded in the research study area were *Striatura milium* (the most numerous species, occurring at 65% of the sites), *Zonitoides arboreus* (the fourth most numerous species, but occurring at 85% of the sites, more than any other species), *Striatura exigua*, and *Discus catskillensis*. Introduced European slugs, *Arion* spp., were also recorded at a few locations.

Amphibians and reptiles

A survey of amphibians and reptiles was conducted along a 0.5-mile-wide corridor stretching from Legion Lake to Hurricane River, and nearby areas, over 24 days between April and August 1990 (Premo and Davis 1990). The researchers stated that for the Upper Peninsula, "the area surveyed was found to be relatively rich in herpetofauna, both in number of species and number of individuals." Twelve amphibian and 7 reptile species were recorded, representing 73% of the herpetofauna known to occur on the Upper Peninsula. However, the areas that supported the largest number of amphibians contained lakes, ponds, and wetlands, all of which are outside the road corridors. No amphibians or reptiles were found to be abundant in the upland areas in the road corridors. The redbacked salamander, wood frog, mink frog, American toad, and red-bellied snake were found to be common in portions of the upland area surveyed. Other species recorded in small numbers on the uplands included gray tree frog, spring peeper, eastern garter snake, and smooth green snake.

Several additional species were not collected in the survey but may be present in the area. In the uplands of the road corridors, the ring-necked snake and fox snake are likely present. The mudpuppy, pickerel frog, and Blanding's turtle may also exist in the area.

Birds

A bird survey was conducted along a 0.5-mile-wide corridor from Legion Lake to the mouth of the Hurricane River, beginning in early May and ending in mid-July 1990 (Dow 1990). A total of 77 species were recorded in the area.[3] In the hardwoods forest that was sampled, some of the most common species observed were red-eyed vireo, ovenbird, blackthroated green warbler, and Nashville warbler. Twelve species were confirmed to be breeding, including yellow-bellied sapsucker, brown creeper, hermit thrush, red-eyed vireo, ovenbird, chipping sparrow, and ruffed grouse. In the pine barrens, the most abundant species observed were black-throated green warbler, chipping sparrow, yellow-rumped warbler, hermit thrush, Nashville warbler, and ovenbird. Six species were confirmed to be breeding in this habitat type: brown creeper, hermit thrush, yellow-rumped warbler, black-throated green warbler, chipping sparrow, and rose-breasted grosbeak.

Dow also detected differences in the diversity and density of species between the shoreline zone and the inland buffer zone: a higher number of species (39) was observed in the inland buffer zone than in the shoreline zone, and 32 species occurred more frequently in the inland buffer zone. Some species were ubiquitous in both areas, such as the red-eyed vireo and ovenbird. Several species occurred statistically more frequently in the inland buffer zone, including the yellow-bellied sapsucker, least flycatcher, common raven, veery, chestnut-sided warbler, black-throated blue warbler, yellow-rumped warbler, American redstart, and scarlet tanager. Conversely, only the black-throated green warbler occurred significantly more frequently within the shoreline zone. These differences between the two zones may be explained by differences in human use of the areas: logging has occurred within the last few years and is continuing in the inland buffer zone, while logging has not occurred in the shoreline zone for many years. As a result, the inland buffer zone has more openings, low cover of different age classes, and forest edge, which in turn provides more niches for birds and thus a greater diversity of species than the more mature, even-aged forests in the shoreline zone.

Some hunting of ruffed grouse, woodcock, and waterfowl occurs in Pictured Rocks National Lakeshore. However, hunting is believed to have a very small effect on the species and populations levels along the road corridors.

Mammals

The Michigan Biological Station estimated that 54 species of mammals occur in Pictured Rocks National Lakeshore (NPS 1980). Over 35 species have actually been observed, or signs of their presence recorded in the park. Some of the most common mammals include deer mice, red-backed vole, eastern chipmunk, red squirrel, snowshoe hare, skunk, raccoon, black bear, coyote, pine marten, and white-tailed deer.

Two studies on the park's mammals were recently completed—one focused on small mammals and threatened and endangered species occurring

---

3. *An additional 34 species were identified in other portions of the park outside the research study area.*

along the shoreline zone corridor and on surrounding lands, and the other on furbearers throughout the park.

Myers and Svendsen (1990) observed and trapped small mammals from Legion Lake to just west of Grand Marais and other park areas from September 15-31, 1990. Altogether, nine rodent species, four shrew species, and one carnivore were captured. Deer mouse the most abundant small mammal captured, and was collected in all of the habitat types sampled. Short-tailed shrews, masked shrews, and red-backed voles were also numerous.

The mixed northern hardwoods and the stand of white birch near the Twelvemile Beach campground were found to support the greatest diversity of mammals (although this may be partly because these areas were more intensively surveyed than other areas). The deer mouse was the most common mammal collected in the maple-beech, mixed northern hardwoods, and birch forest habitat types. Other common species included short-tailed shrew, masked shrew, red-backed vole, and red squirrel. Southern flying squirrel, least chipmunk, eastern chipmunk, pygmy shrew, and short-tailed weasel were also numerous.

Myers and Svendsen only cursorily surveyed the pine forests at the eastern end of the shoreline zone corridor, near the junction of H-58 and the road to the Twelvemile Beach campground. Species that were recorded included deer mouse, eastern chipmunk, redbacked vole, and red squirrel.

It was also noted that numerous least chipmunks were seen on the Kingston Plains. The meadow jumping mouse was found wherever there were patches of grass in the area sampled, and also may be present on the Kingston Plains.

White-tailed deer is the most abundant large mammal in the park; they are occasionally seen in the project area. The park deer population is estimated at roughly 10 deer per square mile in the summer, which is low for the Upper Peninsula (Michigan DNR, Minzey, pers. comm.). Although most deer herds on the eastern half of the Upper Peninsula have been increasing, there are few year-round, resident deer in the park because of the high snow depth; most of the deer head south toward Shingleton. The park deer herd has been declining, due in part to the cessation of logging and suppression of fires in the area (NPS 1980). These actions have diminished the quality and quantity of browse for the deer. Additionally, in the 1970s, the National Park Service terminated a supplemental feeding program in Beaver Basin, which for many years maintained the population at an abnormally high level.

The only other ungulate present in the vicinity of the park is moose. Although moose have been rare on the Upper Peninsula, they are increasing in number. The majority of the moose are currently found from about 5 miles south of M-28 to the Lake Superior shoreline. About 10 moose are believed to use the park as part of their home range, and some of these moose may use the project area (Michigan DNR, Minzey, pers. comm.). All indications are that the moose population will likely increase along the northern tier of the Upper Peninsula, including the park, over the next several years. Eventually, a relatively large number of moose could use parts of the park as parts of their home ranges, possibly reaching a density of one moose per square mile (Michigan DNR, Minzey, pers. comm.).

For his Ph.D. thesis, Daues (1990) studied the occurrence, movements, and use of the park by furbearers from January 1989 through August 1990. He obtained his information through winter trail counts, counts of tracks by scent stations, observations of other field signs, aerial surveys for beaver and muskrat, examinations of harvest records, and interviews of people knowledgeable about large mammals in the area.

Daues found that the northern hardwoods were important for denning and hunting. The largest number of furbearer trails and tracks occurred in this habitat type. However, the coniferous forests and lowlands outside the road corridors were also important for furbearers because of the high densities of prey species such as snowshoe hare and other small mammals found there. Daues also noted that most furbearer species residing in the shoreline zone spend much of their time in the adjacent inland buffer zone.

Daues recorded data on 15 furbearer species in the park. Coyote was found to be the most common furbearer, followed by pine marten, river otter, mink, long-tailed weasel, and red fox. Other common species included the striped skunk, raccoon, black bear, and beaver. Less common species recorded in the park included the eastern timber wolf, bobcat, fisher, short-tailed weasel, and muskrat.

Most of the above species would be expected to occur along the road corridor, with the exceptions of river otter, beaver, and muskrat, all of which usually are found by water. Coyotes generally inhabit semi-open country with a mixture of woods, brushland, and open grassland, and probably can be found throughout the area. Mink occurred more frequently in lowland areas, but occasionally may be seen near the road corridor. Daues found northern hardwoods, mixed northern hardwoods, and coniferous forests in the park to be suitable for long-tailed weasel, although they occurred more often in more mature northern hardwoods. The striped skunk is common to abundant in the park and is probably common along the road corridors, most often in hardwoods. This species is present in various habitats but prefers open or forest edge zones. The creation of forest openings, timber harvesting, and other habitat disturbances favor skunks. Raccoons are also probably common along the shoreline and IBZ corridors. They prefer hardwoods with cavities and interspersed riparian zones. Black bears occur along the road corridors, although the best bear habitat is in the Beaver Basin lowlands. They prefer mixed deciduous/coniferous forest interspersed with marshes and swamps. Bears feed on nut-bearing beech trees along the road corridors and on blueberries in Kingston Plains.

Predators use the corridor areas on an infrequent basis. Daues found limited evidence of red fox in northern hardwoods and coniferous forest in the park. Red fox prefers fields and brush areas but may be found in almost any habitat. Bobcat tracks were observed by Myers and Svendsen (1990) in maple/beech forest and in mixed northern hardwoods and conifers outside the shoreline zone corridor, and Daues reported a few tracks in hardwoods and lowland habitats. This species uses a wide variety of forest habitats and could be found almost anywhere along the corridors. The fisher is another rare species in the park. Daues found some tracks of fishers in a variety of habitats. They use mature hardwoods as denning sites and as hunting areas for porcupine. Their preferred habitats include northern hardwoods and conifers but they may also do well in second-growth forests in the inland buffer zone. Short-tailed weasels also have diverse habitat preferences. They tend to avoid dense forests and prefer successional or forest edge habitats. They are found more often in brush areas, especially along streams and marshes outside the area of the road corridors. Two other predators, pine marten and gray wolf, are discussed in the next section.

Hunting of black bear, deer, woodcock, ruffed grouse, and waterfowl is believed to have a very small effect on the species and population levels along the proposed road corridors.

Threatened and Endangered Species

In 1991 and 1992, the National Park Service informally consulted with the U.S. Fish and Wildlife Service and the state of Michigan on the presence of state and federally listed threatened and endangered species, federal candidates, and state species of concern in the vicinity of the shoreline zone corridor. The U.S. Fish and Wildlife Service identified three federally endangered species (piping plover, peregrine falcon, eastern timber wolf), two threatened species (bald eagle, Pitcher's thistle), and one candidate species (auricled twayblade) as potentially present near the corridor or its vicinity. None of the alternative road corridors is within any designated or proposed critical habitat for federally listed species.[4]

Plants

Surveys were conducted for state and federally listed vascular plants and bryophytes along the shoreline zone corridor and adjacent areas (NPS 1992; Glime et al. 1991). No federally listed plants were identified in these field surveys. Pitcher's thistle *(Cirsium pitcheri)*, a federally threatened species, is common on the Grand Sable Dunes in the eastern part of the park. This species grows on sand dunes and beaches, which do not occur along the alternative road corridors. McEachern (1992) found the park's population to be healthy and reproducing, occurring throughout the Grand Sable Dune field. She further noted that potential habitat for the plant is present throughout this system. Auricled twayblade *(Listera auriculata)*, a 3C federal candidate species, is found in wet woods and thickets, and is known to be

---

4. It should be noted that these surveys focused on the proposed road alignments within the park boundaries. Surveys for threatened and endangered species have not been done along H-58, in Kingston Plains, and outside the park boundaries.

present in the park. No individuals were observed along the shoreline zone corridor.

One state-threatened species was observed in the inland buffer zone. Flattened oats *(Danthonia compressa)* grows along old logging roads, which provide relatively large gaps for light in the forest canopy (NPS 1992, Loope 1992). This species appears to require bare soil to establish. On the nearby Hiawatha National Forest it grows on sandy soil with a high water table along roads and old two-track roads where soil has been scarified. Although flattened oats grow in or near the IBZ corridor, it would be possible to route the road to avoid the plant.

Two other state-threatened species may exist along the alternative road corridors, although they have not been recorded. New England sedge *(Carex novae-angliae)* is found in moist hardwood understories, and may be present along the Beaver Basin Rim. Dwarf bilberry *(Vaccinium caespitosum)* is occasionally found in the park on exposed rock benches and dry soils on old railroad grades. It may be in the Kingston Plains, where there are many old logging era railroad grades. Park biologists intensively searched for these species along the IBZ corridor in the summer of 1992, but none were found.

Animals

Between 1990 and 1992 biologists conducted surveys to identify state and federally listed invertebrates, reptiles and amphibians, birds, and mammals in and near the shoreline zone corridor (Pearce, Nielsen, and Rogers 1992; Premo and Davis 1990; Dow 1991; Myers and Svendsen 1990). No federally listed invertebrates, amphibians, or reptiles have been identified in the area, and no federally listed birds or mammals are known to reside there.

The bald eagle *(Haliaeetus leucocephalus)* was the only federally and state threatened bird species observed near the alternative road corridors by Dow (1991). Two active bald eagle nests are located in the park, one near the south end of Beaver Lake and the other on the southeast end of Sable Lake. No riparian or aquatic environments occur along the road corridors, which eagles would be expected to use.

The federally and state endangered piping plover *(Charadrius melodus)* is known to use the park and adjacent lands. However, the beach habitat that this bird depends on is not in the road corridors, and no piping plovers are known to use the project area. Nordstrom (1990) observed that, with the possible exception of Sand Point Beach, none of the beaches in the park had suitable habitat to support the reestablishment of piping plovers—physical conditions, existing human use, and/or availability of food are unsuitable for the birds. No piping plovers are known to be currently nesting in the park, although in the recent past they did nest on parkland within Grand Marais. The plovers probably occasionally feed along the park beaches, including Twelvemile Beach, and they are also nesting outside the park in Grand Marais.

The federally and state endangered peregrine falcon *(Falco peregrinus)* also does not reside in the project area. Captive-bred peregrines were released in Pictured Rocks National Lakeshore in 1989 and 1991 and also on nearby Grand Island in the Hiawatha National Forest in 1992. Several were observed along the cliffs during June 1993. In 1994 a pair of peregrines nesting in the Grand Portal area raised two young to fledgling stage. The alternative road corridors are a minimum of 6 miles from the cliff habitat where peregrines nested and have historically had aeries.

One bird species that is a federal category 2 candidate for listing (and a state endangered species) may occur in the area, although it has not been recorded as being present. The loggerhead shrike *(Lanius ludovicianus)* could occur at the edge of coniferous forest and open areas, and in logged portions near the alternative road corridors.

Ares and Svendsen (1990) did not trap, observe individuals, or identify signs of any federally listed mammal species along the alternative road corridors. However, the endangered eastern timber wolf *(Canis* lupus*)* could potentially occur here.

Eastern timber wolf

Wolves once were abundant throughout upper Michigan, including the park. However, with European settlement wolves began to disappear and by 1959 had been extirpated from most of the state. The last known wolf pack on the Upper Peninsula was in the Grand Marais area. This pack covered a large territory, which probably included the Pictured Rocks area (Michigan DNR, Arrows, pers. comm.).

A few wolves have continued to exist on the Upper Peninsula, probably immigrating into the area from Ontario, Canada, and Wisconsin. At least six wolves were estimated to be on the peninsula during the early 1970s (Henderickson et al. 1975, cited by USFWS 1992). Between 1960 and 1986, 16 wolf carcasses were recorded on the Upper Peninsula (Theil and Arrows 1988, cited by Daues 1991). However, wolf numbers are increasing: spring prebreeding surveys conducted by the Michigan Department of Natural Resources recorded 17 wolves on the Upper Peninsula in 1991, 21 in 1992, 30 in 1993, 57 in 1994, and 80 animals in 1995. Most sightings of wolves have been in the west Upper Peninsula, although there are a growing number in the east as well. It is likely that wolf numbers will continue to increase on the Upper Peninsula.

Myers and Svendsen (1990) found no records of wolves permanently living in Pictured Rocks National Lakeshore, or in the vicinity of the alternative road corridors. However, Daues (1990) observed one set of wolf tracks on the southern shore of Beaver Lake on February 25, 1989, and two sets of wolf tracks in the Hiawatha National Forest about 16 miles (24 kilometres) south of the proposed road corridor on March 28, 1990. A wolf was reported to the Michigan Department of Natural Resources crossing H-58, south of Kingston Lake, in July 1990. In the winter and spring of 1993, the Michigan Department of Natural Resources received reports of wolf tracks or scat at four locations in or near the park, including Grand Sable Lake, Chapel Basin (two sightings), Miners River area, and inland from Twelvemile Beach. A wolf dropping also was collected on Twelvemile Beach. Additional sightings have been reported in the general area to the Michigan Department of Natural Resources.

More recent data indicates that wolves are now in the area and are making use of the park. In the winter of 1994–95, the Michigan Department of Natural Resources verified that two wolves were using the western end of the park and surrounding lands and one wolf the east end of the park as part of their home ranges (Michigan DNR, Minzey, pers. comm.). Additional wolves are believed to be south of the park. It is not clear, however, what the range of these wolves is, or whether breeding packs are using the park.

It seems likely that the park could eventually be used as part of a pack territory, if not already. However, Pictured Rocks National Lakeshore is not large enough to support a wolf pack by itself. A wolf pack's territory ranges from 20 to 214 square miles or more (USFWS 1992). In the upper Midwest, including the Pictured Rocks area, 100 square miles is estimated to be about the average amount of land needed to support a colonizing pack (USFWS, Mech, pers. comm.; Mech 1986; Fritts and Mech 1981; Mech 1973). Also, by itself Pictured Rocks National Lakeshore would not provide sufficient deer, moose, or beaver—the primary prey of wolves—to support a wolf pack. As noted earlier, the park supports a low deer population compared to the rest of the peninsula. The park is expected to support increasing numbers of moose in future. But moose are a formidable opponent for wolves and will not be a common prey species. There are areas in the park that support relatively high beaver densities, including Beaver Basin, and the Sevenmile Creek and Sable Lake areas. But there are no beaver in the project area. Daues (1991) also found the park overall beaver colony density to be 0.21/square kilometer, which is on the low end of the average colony density in North America.

In 1992 the U.S. Fish and Wildlife Service revised its recovery plan for the eastern timber wolf. The primary objective of the plan is to "maintain and reestablish viable populations of the eastern timber wolf in as much of its former range as is feasible". One of the areas to be investigated for reestablishment of wolves is Michigan's 16,538-square-mile Upper Peninsula. The best potential wolf habitat is probably on the Lake Michigan watershed side of the Upper Peninsula, which has a large deer population compared to the remainder of the peninsula (Michigan DNR, Arrows, pers. comm.). On the Lake Superior watershed side of the peninsula, the Fish and Wildlife Service has set no wolf population goals for the park. However, south of the park the Fish and Wildlife Service has set a planning goal of one wolf pack, consisting of six wolves, becoming established in the Hiawatha National Forest and Seney National Wildlife Refuge by the year 2000 (USFWS 1992). Since there are no plans to translocate wolves on the peninsula, if this goal is to be achieved wolves will have to naturally reestablish themselves in the area.[5]

The Michigan Department of Natural Resources is currently developing a recovery and management plan for the wolf in Michigan. The purpose of this plan is to "ensure the long-term survival of a self-sustaining wild wolf popula-tion in Michigan". This plan provides population restoration goals and recovery strategies for the Upper Peninsula, although it does not provide specific goals and strategies for wolves using the park and vicinity. It also does not recommend wolf reintroductions on the Upper Peninsula.

Other state-listed species

Several state-listed animal species probably occur along the road corridors, although only a few species have actually been recorded as present. One state-threatened species that Daues (1991) observed to be relatively common in the park (although Myers and Svendsen did not record it) was pine marten (*Martes americana*). This species was thought to have become extinct in Michigan by 1940. But subsequent reintroductions of marten on the Upper Peninsula in the 1950s, 1960s, and 1980s appear to have succeeded. Indeed, prints of marten were among the most common furbearer tracks observed by Daues in the park and were found in most of the areas he surveyed. However, they appear to prefer mature upland northern hardwoods, probably due to suitable denning areas there. Myers and Svendsen believe marten are probably present in low numbers along the shoreline zone corridor, especially in mixed northern hardwoods/coniferous forest, maple/beech forest, and birch forest habitats. They believe that the most important marten habitat is probably the mixed conifers and northern hardwoods along the Beaver Basin Rim where red squirrels are abundant.

Several invertebrates of state special concern may exist along the alternative road corridors. Carolina mantleslug (*Philomycus carolinianus*) was recorded by Pearce, Nielsen and Rogers (1992) at Bad Gully, at the top of the escarpment above Beaver Basin in beechsugar maple forest. It likely exists in low numbers throughout the area's beech-maple forest. Other species that may occur along the alternative road corridors but that have not been observed include early hairstreak *(Erora laeta)*—likely occurs in the western portion of the hardwoods; Doll's merolonche *(Merolunche dolli)*—may occur in the area's open pine barrens; and hoary comma *(Polygonia gracilis)*—may occur along moist sunny trails and roads. There is also a remote possibility that the state-threatened northern blue butterfly *(Lycaeides idas nabokovi) is* present on Kingston Plains, although its food plant, dwarf bilberry, has not been found there.

---

5. *A wolf pack was verified in the east unit of the Hiawatha National Forest in the winter of 1992-93. Wolves were also verified in the west unit in the winter of 1993-94 (Michigan DNR, Arrows, pers. comm.).*

Besides the bald eagle, several state-listed bird species may reside in or along the alternative road corridors, or migrate through the area. The osprey *(Pandioin haliaetus)*, a state-threatened species, may pass through the area, and was seen during the bird survey on Beaver Lake outside the alternative road corridors. However, this bird probably does not make use of the area due to the absence of lake and river habitat. Although there are no records of their presence in the park, the state-threatened lark sparrow *(Chondestes grammacus)* may reside at the edge of coniferous forest and open areas, and the threatened yellow-throated warbler *(Dendroica dominica)* may reside in coniferous forests in the area. Other species that may migrate through the area include the state-endangered short-eared owl *(Asio flammeus)*, and the threatened red-shouldered hawk *(Buteo lineatus)*, long-eared owl *(Asio otus)*, and prairie warbler *(Dendroica discolor)*. The state-endangered merlin *(Falco columbarius)* has been recorded at Sand Point outside the road corridors. In 1993, four young fledged from a nest. Two species of special concern may also migrate through the area—Cooper's hawk *(Accipiter cooperi)* and northern harrier (Circus *cyaneus).*

Vegetation

ANALYSIS

Construction of the proposed main road would result in a permanent loss of approximately 82 acres of forest vegetation, including 69 acres of mixed northern hardwoods forest and 13 acres of pine woodlands.[6] Construction of the two Beaver Basin overlooks, including the access roads and parking areas, would result in the loss of an additional 5 acres of mixed northern hardwoods forest. For the purposes of vista clearing, trees on another 13 acres would be pruned and selectively cut on a periodic basis to provide visitors with a view of the basin and Lake Superior. Tall trees at the overlooks would be selectively removed from the slope, but most of the forest vegetation here should not be affected.

The complete removal of forest vegetation along the road corridor and at the overlooks would have many direct and indirect effects. The loss of the canopy would increase light in the area and change the microclimate of what was an interior forest. Although the trees along the edge of the road would grow and branch out with time, the width of the road's construction limits would probably prevent the canopy from completely closing at least within the next 15 years. The forest along the roadside would become an edge or disturbed community. Some shade-tolerant understorey plants would probably become less abundant, while other species that require more light would increase in abundance. Shrubs, young trees, and other understory and ground vegetation would probably become denser along the forest edge.

---

6. *This assumes a 13-mile-long alignment, an average 52-foot-wide construction limit (ranging between 46 and 54 feet), and proper implementation of all the mitigating measures described in the alternative.*

Construction of the road would result in a modified, artificially maintained vegetative community along the roadsides. Of the total 82 acres of forest vegetation that would be cleared for the main road, approximately 47 acres within the road construction limits (i.e., shoulders, drainage ditch, and adjacent slopes) would be revegetated with grasses and forbs that would grow in the new cleared roadside environment. About another 2 acres would be revegetated along the two overlook access roads. Some hardy grasses that can withstand periodic mowing would be planted in the stabilized turf shoulders. Most of these species would be native grasses, with perhaps one or two species such as red fescue that are not native to the area. However, these species would not spread outside the roadside.

The new road would also increase the potential for the introduction of exotic plants in the project area, particularly along the roadsides. Although the mitigation measures listed in the alternative would help prevent their spread, plants such as spotted knapweed and St. John's Wort (which already exist in the project area) would probably still establish along the roads and possibly become abundant. Seeds from some exotic plants, such as periwinkle, could also disperse into the understory of the adjacent closed canopy forest.

CONCLUSION

Approximately 87 acres of mixed northern hardwoods and pine woodland would be affected, and some tall trees on another 13 acres would be selectively cut for vista improvement purposes at the two overlooks. See Table 8 for summary of vegetation affected under alternative B. The existing forest canopy would be lost, which would increase the abundance of some native plants and decrease others along the road edges. There would also be an increased potential for the spread of exotic plant species along the roadsides. Overall, however, alternative B would have a minor affect on the entire park vegetation, affecting about 0.1% of the existing forest cover.

Table 8. Summary of vegetation acreage affected—Alternative B

	Total vegetation disturbed	Vegetation lost (pavement)	Revegetated area
Main Road	82 acres	35 acres	47 acres
Spur Road and Overlooks	5 acres	3 acres	2 acres
Total	87 acres *	38 acres	49 acres

*Tall trees on another 13 acres would be selectively cut to provide visitors with a view at the two overlooks.

Wildlife

ANALYSIS

Construction and use of the proposed road would have a variety of effects on the invertebrates, amphibians, reptiles, birds, and mammals in the area. The road, spur roads, parking areas, and overlooks would result in the loss or alteration of approximately 87 acres of forest habitat. None of the resource studies conducted for this project indicated that the area is of special importance for nesting, breeding, or foraging, nor are there any migration routes known to cross through the area. Clearing within the construction limits would result in an initial loss of forage and shelter for ground- and tree-dwelling species. Some of this forage and shelter would be reestablished over time. However, the relatively narrow width of the construction limits would be too small to affect species other than perhaps some insects. There would be no appreciable change in species' populations or distributions. Most local wildlife populations would compensate for the roads, shifting their home ranges. Placing the road away from the Rim would also help minimize impacts on sensitive species such as the pine martin (Myers and Svendsen 1991).

Providing an opening in the forest and a roadside habitat would cause some seasonal changes in wildlife use of the area, although there would probably be a negligible change in populations. For instance, in the summer white-tailed deer may be attracted by the roadside grasses and make more use of the project area. The road corridor habitat and adjacent forest edge would attract species that use edge ecotones for feeding and nesting. Some animals, such as skunk, coyote and scavengers like vultures may find addi-tional food sources (roadkills) along the roadside.

During the construction phase there would be a short-term disturbance and displacement of animals. Most birds, mammals, reptiles, and amphibians would avoid sites that are being cleared, graded, or actively used by people. Noise from construction equipment would likewise also probably displace animals. Many of these animals most likely would return after construction activities cease and the vegetative cover is restored. Some animals, primarily invertebrates, would not be able to move out of the construction area and would be killed.

Some wildlife mortality would occur when the road is opened to public use. Roadkills would result from wildlife being hit by vehicles, particularly in areas with short sight distances. However, the relatively low projected use of the road and the low posted speed would minimize this mortality. Although the road would increase access for hunters, hunting would not be expected to increase due to the lack of desired game in the project area and a hunting prohibition within 100 feet of the road. The only exception would be that disabled hunters would be permitted to shoot from their vehicles under applicable state and federal laws. Some hunting that occurs along the rim would be eliminated due to the presence of the road. The road could increase the potential for wildlife poaching, but with use of the road by the public and law enforcement patrols, poaching would probably be negligible. Hunting pressure and the potential for poaching may actually decline in Beaver Basin with closure of old existing roads and trails, which would be eliminated upon construction of the new road.

Large numbers of snowmobiles may also use the new road. The effects of snowmobiles on wildlife would vary depending on the species in the project area. Relatively few animals would be present during the winter. The species in the project area in the winter most likely to be affected would be furbearers and small mammals such as rabbits. Some animals may flee areas adjacent to the road or be temporarily displaced (although these animals may desert the area due to the presence of automobiles). Research has shown that snowshoe hares avoid snowmobile trails (Neumann and Merriam 1972). Some mortality of small mammals may occur, such as mice and voles, when snow is compacted by snowmobiles (Bury 1979; Wall and Wright 1977, as cited in Kuss et al. 1990). However, furbearers and small browsing animals may take advantage of the compacted snow and use the road as travel routes when snowmobiles are not present. Because snowmobiles would be restricted to the road, harassment of wildlife in other parts of the park would not be expected. The heavily wooded terrain along the road, would also curtail snowmobiles driving off the road.

The new Beaver Basin and Sevenmile Creek viewpoints would create new ecotones and would have a minor affect on the behavior, populations, and distributions of resident species in a limited area. A few small mammals, such as rodents, may be displaced by construction of parking lots. Some nesting birds and mammals sensitive to the presence of people would avoid the areas. On the other hand, animals such as chipmunks, red squirrels and blue jays could be attracted to the viewpoints if people eat meals or feed animals there. Concentrations of these species could increase on a seasonal basis.

Concerns have been expressed that constructing a Beaver Basin Rim road would contribute to habitat fragmentation, which would affect wildlife adversely.[7] Habitat fragmentation can be defined as "the subdivision of once large and continuous tracts of habitat into smaller patches" (Rosenfield et al. 1991). Roads are clearly one source of habitat fragmentation, along with agriculture, urbanization, and other service right-of-ways. Habitat fragmentation and creation of edge habitat have been shown to change songbird diversity and abundance, reduce bird nesting success (due to increases in nest predation and parasitism), and affect the abundance and behavior of some mammals, reptiles, and amphibians. However, some tolerant species, such as the mourning dove, flicker, woodcock, white-crowned sparrow, raven, and coyote, could take advantage of the new ecotone and increase their use of this area. Roads can also serve as dispersal corridors for some species, providing a grassy corridor through a forest. On the other hand, they can act as dispersal barriers to small mammals, birds, and insects. For instance, small forest mammals have been shown to be reluctant to cross roads more than 60 feet wide from forest edge to edge (Adams and Geis 1981). In general, the smaller the animal, the more likely a road would become a barrier.

---

7. *The following discussion of habitat fragmentation is largely based on a literature review of the potential effects of habitat fragmentation from the Beaver Basin Rim road (Rosenfield et al. 1991).*

However, most studies of habitat fragmentation have mainly been conducted in highly fragmented landscapes, where only remanent forests remain. Many studies have focused on the effects of fragmentation on birds. The mechanisms by which forest fragmentation impacts occur "seem to be complex, variable, and in dispute" (Rosenfield et al. 1991). There is little information on the potential for short- and long-term impacts of habitat fragmentation in the Pictured Rocks area, which is still largely forested. Researchers do not know what constitutes an effective barrier or corridor in this region. It should also be noted that the road would not cause major changes in the arrangement and aerial extent of different types of forest patches, which would make the isolation of habitat patches seem unlikely. Furthermore, the park area and surrounding lands have already been fragmented for many years, by county and state roads, private logging roads, and logging operations. There are already over 2 miles of linear road per square mile in this region and additional developments such as recreational second homes may be built in future. The effects of all these developments on fragmentation of wildlife habitat is unknown.

Constructing a road through Pictured Rocks National Lakeshore would change the pattern and extent of forest edge habitats, which in turn could affect the resident wildlife community. However, as mentioned above, the existing wildlife community has probably already been affected to an unknown extent by previous and existing roads and other human activities. The new road may not have much additional affect on wildlife. In their review of the literature, Rosenfield et al. (1991) noted the following:

> There is substantial evidence that forest fragmentation and the consequent creation of edge habitats may have adverse impacts on the presence, numbers, and/or reproductive success of area-sensitive species within the resultant forest fragments, while benefiting other species (some of which are predators or brood parasites on forest interior species) through the provision of new edge habitats.

At this time it is not possible to forecast which wildlife populations in the project area would benefit and which would be adversely affected by habitat fragmentation resulting from the proposed road.

CONCLUSION

Overall, the proposed road would be expected to have a negligible effect on most of the park wildlife populations; at most, there may be some minor changes in bird and mammal populations, behavior and distributions in the vicinity of the viewpoints. Construction activities would temporarily displace some animals but most would return after construction ceases. There would probably be some increase in mortality, primarily due to construction activities and roadkills, but this would not be expected to adversely affect local populations. The road would increase habitat fragmentation in the project area, which might benefit some species and adversely affect others to an unknown degree.

Recovery of the eastern timber wolf

ANALYSIS

As noted under alternative A, Pictured Rocks National Lakeshore may be a marginal habitat for wolves compared to other portions of the Upper Peninsula. The park does not support a large deer herd nor other prey for wolves. Recreational use is increasing in the area. The existing road density is over 2.3 linear miles per square mile of land. (Wolf populations usually do not sustain themselves in areas where rural roads open to the public have densities exceeding 0.93 linear mile of road per square mile of land.) Adding the 13-mile-long proposed road and closing approximately 3 miles of roads in Beaver Basin would have a net effect of slightly increasing the road density to approximately 2.6-2.7 linear miles of road per square mile. This change would probably not be a major factor in wolf use/nonuse of the park and surrounding areas.

The presence of a Beaver Basin Rim road poses three potential concerns for wolf recovery: (1) the road could result in wolf mortality (either through illegal take or roadkills); (2) it might constitute a barrier to wolf entry into the project area; or (3) it could discourage wolves from using the park as a possible breeding/denning area. The road would provide easier access to people who could deliberately or accidentally kill wolves by shooting, trapping, or snaring. This potential impact could be reduced with education, public outreach programs, careful monitoring, and law enforcement. The lack of desired game in the area and prohibition on hunting within 100 feet of the road would also help reduce the potential for illegal taking.

With regard to roadkills, Fuller [n.d.] noted that vehicular accidents are just one of several causes of human-induced wolf deaths. Field studies in northwestern Wisconsin on the other hand, have identified roads as a significant cause of wolf mortality (Kohn, Shelley et al. 1995). Five wolves are known to have been struck and killed by vehicles on the Upper Peninsula since 1989. These wolves were killed primarily in the extreme east and west end of the peninsula, of which three were killed on paved, high-speed roads (Michigan DNR, Hammill, pers. comm.). Wolf roadkills on the proposed Beaver Basin Rim road are considered unlikely due to the narrow width of the road, slow speed limits, and projected use levels. The effect of any collision would be inconsequential to the wolf population in the area.

The road could potentially be a barrier for wolves traveling through the area. Major highways can be barriers to dispersal. Stebler (1951) observed that wolves were leery of crossing county road H-58 when he studied them in the area in the 1930s and in 1950. In Alaska, Thurber et al. (1994) found wolf use of lands adjacent to roads varied: wolves tended to avoid a high-use oilfield access road on the Kenai Peninsula, but did not wholly avoid lands adjacent to a well-traveled, paved, high-speed state highway. They appeared to be attracted more to secondary gravel roads with limited human use, however. Studies in Wisconsin indicate that in most situations two-lane roads are not an impediment to wolf movement (Wisconsin DNR, Adrian Wydevian, pers. comm.). Other roads already in the region are more likely to be barriers than the park road. Specifically, wolves would probably have to periodically cross state highway M-28, south of the park, if they are to reestablish a territory that includes the project area. The proposed park road would be narrower than M-28, be used by far fewer vehicles (and no commercial trucks), have a slower speed limit (35 mph versus 55 mph), and be seasonally closed—all of which would reduce the likelihood of the park road being a barrier to wolf movements. Wolves could actually use the park road as a travel corridor, particularly during the winter if snowmobiles pack down the snow (although large numbers of snowmobiles might inhibit usage of this route).

There is no evidence that construction of the proposed road would preclude wolf usage of the park for denning. Studies in Wisconsin indicate that roads may discourage the use of areas by wolf packs; wolf packs are rarely found in areas where their territory is crossed by a paved road (Wisconsin DNR, Wydevian, pers. comm.). But regardless of the road, nobody knows whether wolves would use the project area for denning given the changes that would occur in the natural environment. True, wolves need some seclusion but are known to establish dens in a wide variety of site conditions. For instance, wolf dens were found within 0.6 mile of a major highway in Alaska (Thurber et al., 1994). Mech (1995) noted that breeding packs now live less than 56 miles (90 km) from Twin Cities, Minnesota. Historically, wolves probably bred and denned throughout the Upper Peninsula under a variety of conditions. Although wolves probably would not den in close proximity of the proposed road, there is sufficient habitat outside the project area, both inside and outside the park, for them to den and raise pups. Constructing a road in the park should not affect wolf propagation.

Concerns have also been raised about the impact snowmobiles might have on wolves. As noted in the visitor projections (see section "Impacts on Visitor Use and Experience"), there is the potential for a significant number of snowmobiles to use the new road - an estimated 23,600 snowmobiles may make use of the road from mid-December through mid-March in 2013, or about 262 snowmobiles per day. There is no evidence to indicate that these snowmobiles would have a negative affect on the wolves using the project area. Fuller (n.d.) stated that there is no evidence that snowmobiles affect wolf populations, unless they are specifically intending to harass wolves. It would be very difficult to chase a wolf on a snowmobile off the road in the wooded terrain adjacent to the proposed road. (Wolf harassment would be much more likely in open country where snowmobiles can pursue wolves at high speeds.) It is also worth pointing out that the Upper Peninsula is one of the highest snowmobile use areas in the country, but wolves are increasing on the peninsula in spite of this. Indeed, if snowmobiles pack the snow down on the road, wolves may use the road as a travel route during times when snowmobiles are absent.

Finally, it should be remembered that wolves are adaptable animals and have extended their ranges into areas with considerable development. Mech (1995) observed that changing public attitudes toward wolves, increased publicity, and legal protection, have enabled wolves to use areas that have not been used for decades, "… thus demonstrating the wolf's inherent adaptability." Fuller (n.d.) similarly observed that:

Wolf populations seem little affected by snowmobiles, cars, trucks, logging, mining, and other human activities, except as they might facilitate killing by humans. Wolves are adaptable; they enter towns or villages at night, cross four-lane highways and open landscapes, den near logging sites, and frequent open pit mines and garbage dumps.

Conclusion

Based on present information, constructing the proposed road would not likely adversely affect the eastern timber wolf or wolf recovery in the park and on the Upper Peninsula at this point in time. Regardless of the road, it is expected that wolves would continue to be rare in the vicinity of the project area. The presence of the road in the shoreline zone would not prevent wolves from traveling through the park or denning in nearby areas. If sufficient prey are available and human activity is generally low, wolves may make use of the area with or without the road. The road would slightly increase the potential for wolf mortality. However, closing the existing old forest roads and trails that connect with the new road would be a beneficial effect, reducing human access outside the road corridor and the potential for the illegal taking of wolves.

References

Aanen, P., Alberts, W., Becker, G.J., van Bohemen, H.D., Melman, P.J.M., van der Sluijs, J., Veenbaas, G., Verkaar, H.J. and van de Watering, C.F. 1991. *Nature Engineering and Civil Engineering Works*. Pudoc, Wageningen.

Adams, L.W. 1984. Small mammals use of interstate highway medium strip. *J. Applied Ecology*, 21: 175-178.

Adams, L.W. and Geis, A.D. 1983. Effects of roads on small mammals. *J. Applied Ecology*, 20: 403-415.

Agren, C. 1996. Road traffic a key source of nitrogen deposition. *Enviro*, 20: 19-20.

Amaranthus, M.P., Rice, R.M., Barr, N.R. and Ziemer, R.R. 1985. Logging and forest roads related to increased debris slides in southwestern Oregon. *J. Forestry*, April 1985, pp. 229-233.

Amor, R.L. and Stevens, P.L. 1975. Spread of weeds from a roadside into sclerophyll forests at Dartmouth, Australia. *Weed Research*, 16: 111-118.

Amrhein, C. and Strong, J.E. 1990. The effect of de-icing salts on trace metal mobility in roadside soils. *J. Environmental Quality*, 19: 765-772.

Andres, D.L. and Andres, C.J. 1995. Roadside litter and current maintenance waste management practices: are we making any progress? pp. 135-143. In: *Maintenance Management—Proceedings of the Seventh Maintenance Management Conference*, Transportation Research Board/National Research Council, Florida, July 1994.

Andrews, A. 1990. Fragmentation of habitat by roads and utility corridors: a review. *Australian Zoologist*, 26: 130-141.

Angold, P.G. 1992. *The role of buffer zones in conservation of semi-natural habitats*. Ph.D. Thesis. Univ. Southampton, England.

Angold, P.G. 1997. The impact of a road upon adjacent heathland vegetation: effects on plant species composition. *J. Applied Ecology*, 34: 409-417.

Arnold, G.W. and Weeldenburg, J.R. 1990. Factors determining the number and species of birds in road verges in the wheatbelt of Western Australia. *Biological Conservation*, 53: 295-315.

Ash, C.P.J. and Lee, D.L. 1980. Lead, cadmium, copper and iron in earthworms from roadside sites. *Environmental Pollution* (Series A), 22: 59-67.

Atkins, D.P., Trueman, I.C. and Clarke, C.B. 1982. The evolution of lead tolerance by *Festuca rubra* on a motorway verge. *Environmental Pollution* (Series A), 27: 233-241.

Atkinson, R.B. and Cairns, J. 1992. Ecological risks of highways. *Advances in Modern Environmental Toxicology*, 20: 237-262.

Bach, W. 1985. 'Waldsterben': our dying forests—Part III. Forest dieback: extent of damages and control strategies. *Experientia* 41: 1095-1104.

Backhaus, B. and Backhaus, R. 1987. Distribution of range transported lead and cadmium in spruce stands affected by forest decline. *Science Total Environment*, 59: 283-290.

Bakowski, C. and Kosakiewicz, M. 1988. Effects of forest road on Bank Vole and Yellow-neck populations. *Acta Theriologica*, 72: 245-252.

Baur, A. and Baur, B. 1990. Are roads barriers to dispersal in the land snail *Arianta arbustorum*? *Can. J. Zoology*, 68: 613-617.

Beanlands, G.E. and Duinker, P.N. 1984. An ecological framework for environmental impact assessment. *J. Environmental Management*, 18: 267-277.

Beeby, A. 1985. The role of *Helix aspersa* as a major herbivore in the transfer of lead through a polluted ecosystem. *J. Applied Ecology*, 22: 267-275.

Bellis, E.D. and Graves, H.B. 1978. Highway fences as deterrents to vehicle-deer collisions. *Transportation Research Record*, 674: 53-58.

Bellinger, E.G., Jones, A.D. and Tinker, J. 1982. The character and disposal of motorway run-off water. *Water Pollution Control*, 81: 372-390.

Bennett, A.F. 1988. Roadside vegetation: a habitat for mammals at Naringal, southwestern Victoria. *Victorian Naturalist*, 105: 106-113.

Bennett, A.F. 1991. Roads, roadsides and wildlife conservation: a review pp. 99-118. In: Saunders, D.A. and Hobbs, R.J. (eds.). *Nature Conservation 2: The Role of Corridors*. Surrey Beatty & Sons. Chipping Nortan, Australia.

Bickmore, C.J. 1990. Wildlife, roads and rivers. Landscape Design. *J. Landscape Institute*, 190: 54-56.

Black, D. 1993. *The Campaign to save Oxleas Wood. The History and Rationale of the Battle against the East London River Crossing, to Preserve Plumstead and the Open Spaces of Greenwich*. PARC, London.

Boarman, W.I. and Sazaki, M. 1996. Highway mortality in desert tortoises and small vertebrates: success of barrier fences and culverts pp. 169-173. In: Evink, G., Ziegler, D., Garrett, P. and Berry, J. (eds.). *Transportation and Wildlife: Reducing Wildlife Mortality and Improving Wildlife Passageways across Transportation Corridors*, Florida, Federal Highway Administration and Florida Department of Transportation. Conference Report No. FHWA-PD-96-041.

Box, J.D. and Forbes, J.E. 1992. Ecological considerations in the environmental assessment of road proposals. *Highways and Transportation*, April 1992, pp. 16-22.

Boxall, A.B. and Maltby, L. 1995. The characterisation and toxicity of sediment contaminated with road run-off. *Water Research* 29: 2043-2050.

Braun, S. and Fluckiger, W. 1984a. Increased population of the aphid *Aphis pomi* at a motorway. Part 1: Field evaluation. *Environmental Pollution*, 33: 107-120.

Braun, S. and Fluckiger, W. 1984b. Increased population of the Aphid *Aphis pomi* at a motorway. Part 2: The effect of drought and deicing salt. *Environmental Pollution*, 36: 261-270.

Brockie, R. 1960. Road mortality of the hedgehog (*Erinaceus europaeus*) in New Zealand. *Proc. Zoological Society of London*, 134: 505-508.

Brockie, R. 1999. For the love of hedgehogs. *New Zealand Geographic*, 44: 96-109.

Brody, A.J. and Pelton, M.R. 1989. Effects of roads on Black Bear movements in western North Carolina. *Wildlife Society Bulletin*, 17: 5-10.

Brothers, T.S. and Spingarn, A. 1992. Forest fragmentation and alien plant invasion of central Indiana old-growth forests. *Conservation Biology*, 6: 91-100.

Brown, R., Brown, M. and Pesotto, B. 1986. Birds killed on some secondary roads in Western Australia. *Corella*, 10: 118-122.

Brown, T. 1993. Resource Management Act has major impact on roading projects. *Local Authority Engineering in New Zealand*, 9: 15-22.

Burgaz, S., Erdema, O., Karahalil, B. and Karakaya, A.E. 1998. Cytogenetic biomonitoring of workers exposed to bitumen fumes. *Mutation Research*, 419: 123-130.

Burke, R.C. and Sherburne, J.A. 1982. Monitoring wildlife populations and activity along I-95 in Northern Maine before, during and after construction. *Transportation Research Record*, 859: 1-8.

Burns, J.W. 1972. Some effects of logging and associated road construction on Northern California streams. *Trans. American Fisheries Society*, 101: 1-17.

Byron, H. 1999. *Biodiversity and Environmental Impact Assessment of Road Schemes. Draft Guidance on a Systematic Approach*. London, Imperial College of Science, Technology and Medicine, Environmental Policy and Management Group, T.H. Huxley School of Environment, Earth Sciences and Engineering.

Byron, H. 2000. Biodiversity Impact. In: *Biodiversity and environmental impact assessment: a good practice guide for road schemes*. The RSPB, WWF-UK, English Nature and Wildlife Trusts, Sandy.

Canters, K.J. and Cuperus, R. 1997. Assessing fragmentation of bird and mammal habitats due to roads and traffic in transport regions pp. 160-170. In: K. Canters (ed.). Habitat Fragmentation and Infrastructure. *Proc. International Conference 'Habitat fragmentation, Infrastructure and the Role of Ecological Engineering'*, Maastricht, The Hague. Delft, Ministry of Transport, Public Works and Water Management.

Canters, K.J., Piepers, A. and Hendriks-Heersma, D. (eds.) 1997. *Habitat Fragmentation and Infrastructure and the Role of Ecological Engineering. Proc. International Conference*, 17-21 September 1995, Maastricht, the Hague, The Netherlands. Delft, Ministry of Transport, Public Works and Waste Manage-ment, Road and Hydraulic Engineering Division.

Carr, M.H., Zwick, P.D., Hoctor, T., Harrell, W., Goethals, A. and Benedict, M. 1998 pp. 68-77. In: Evink, G.L., Garrett, P., Zeigler, D. and Berry, J. (eds.) *Proc. International Conference on Wildlife Ecology and Transportation*. Using GIS for identifying the interface between ecological greenways and roadways systems at the state and Sub-state scales. FL-ER-69-98. Florida Department of Transportation, Tallahassee, Florida.

Carroll, A.B. 1995. *Business and Society: Ethics and Stakeholder Management*. Southwestern Publishing Co., Cincinnati.

Case, R.M. 1978. Interstate highway road-killed animals: a data source for biologists. *Wildlife Society Bulletin*, 6: 8-13.

Cheshire Ecological Services. 1995. Cheshire Roadside verge survey. Grebe house, Rease heath, Nantwich, Cheshire, U.K.

Chmiel, K.M. and Harrison, R.M. 1981. Lead content of small mammals at a roadside site in relation to the pathway of exposure. *Science Total Environment*, 17: 145-154.

Chomitz, K.M. and Gray, D. 1996. Roads, land use, and deforestaton: a spatial model applied to Belize. *World Bank Economic Review*, 10: 487-512.

Clevenger, A.P. and Waltho, N. 2000. Factors influencing the effectiveness of wildlife underpasses in Banff National park, Alberta, Canada. *Conservation Biology*, 14: 47-56.

Clevenger, A.P., Chruszoz, B. and Gunson, K. 2001. Drawings culverts as habitat linkages and factors affecting passage by mammals. Journal of Applied Ecology, 38, 1340–1349.

Clifford, H.T. 1959. Seed dispersal by motor vehicles. *J. Ecology*, 47: 311-315.
Clinnick, P.F. 1985. Buffer strip management in forest operations: a review. *Australian Forestry*, 48: 34-35.
Cook, K.E. and Daggett, P-M. 1995. *Highway Roadkill, Safety, and Associated Issues of Safety and Impact on Highway Ecotones*. Washington, Transportation Research Board.
Coulson, G. 1982. Road-kills of macropods on a section of highway in central Victoria. *Australian Wildlife Research*, 9: 21-26.
Cowie, I.D. and Werner, P.A. 1993. Alien plant species invasive in Kakadu National Park, Tropical Northern Australia. *Biological Conservation* 63: 127-135.
Cramer, G.H. II and Hopkins, W.C. Jr. 1982. Effects of dredged highway construction on water quality in a Louisiana Wetland. *Transportation Research Record*, 896: 47-51.
Crawley, M.J. 1989. Chance and timing in biological invasions pp. 407-423. In: Drake, J.A., Mooney, H.A., di Castri, F., Groves, R.H., Kruger, F.J., Rejmanek, M. and Williamson, M. (eds.). *Biological Invasions—A Global Perspective*, SCOPE & John Wiley & Sons, NY.
Criley, M. 2000. From gravel to pavement—the impacts of upgrading. *The Road RIPorter*, July/August, 5: 12-13.
Criley, M. 2001. Understanding the new National Forest System road management strategy. *The Road-RIPorter*, March/April, 6: 4-5.
Crombie, R.I. and Heyer, W.R. 1983. *Leptodactylus longirostris* (Anura: Leptodactylidae): advertisement call, tadpole, ecological and distributional notes. *Revue Brasilian Biology*, 43: 291-296.
Crowe, S. 1960. *The Landscape of Roads*. Architectural Press, London.
Cuperus, R., Canters, K.J. and Piepers, A.A.G. 1996. Ecological compensation of the impact of a road. Preliminary method for the A50 road link (Eindhoven-Oss, The Netherlands). *Ecological Engineering*, 7: 327-349.
Cuperus, R., Canters, K.J., Udo de Haes, H.A. and Friedman, D.S. 1999. Guidelines for ecological compensation associated with highways. *Biological Conservation*, 90: 41-51.
Curzydlo, J. 1985. Effects of forest and roadside shelterbelts on the spread of toxic components of car exhausts. *Sylwan*, 129: 21-30.
Dalrymple, H.H. and Reichenbach, N.H. 1984. Management of an endangered species of snake in Ohio, USA. *Biological Conservation*, 30: 195-200.
Davies, J.M., Roper, T.J. and Shepherdson, D.J. 1987. Seasonal distribution of road kills in the European Badger (*Meles meles*). *J. Zool.* (London) 211: 525-529.
Davis, J.R. and George, J.J. 1987. Benthic invertebrates as indicators of urban and motorway discharges. *Science Total Environment* 59: 291-302.
Davison, A.W. 1971. The effects of de-icing salt on roadside verges. I, soil and plant analysis. *J. Applied Ecology*, 8: 555-561.
de Hamel, B. 1976. *Roads and the Environment*. HMSO, London.
Dennis, R.L.H. 1986. Motorways and cross-movements. An insect's 'mental map' of the M56 in Cheshire. *AES Bulletin*, 45: 228-243.
Department of Transport (1992). *Assessing the environmental impact of road schemes*. London, HMSO.
Deroanne-Bauvin, J.E.D. and Impens, R. 1987. Monitoring lead deposition near highways: a ten-year study. *Science Total Environment*, 59: 257-266.
Detwyler, T.R. 1971. *Man's Impact on Environment*. McGraw Hill, NY.

Diaz, J.A., Carbonell, R., Virgos, E., Santos, T. and Telleria, J.L. 2000. Effects of forest fragmentation on the distribution of the lizard *Psammodromus algirus*. *Animal Conservation*, 3: 235-240.

Dickerson, L. 1939. The problem of wildlife destruction by automobile traffic. *J. Wildlife Management*, 3: 104-116.

Dickson, K.L. 1986. Neglected and forgotten contaminants affecting aquatic life. *Environmental Toxicology and Chemistry*, 5: 939-940.

Disecker, E.G. and Richardson, E.C. 1961. Roadside sediment production and control. *Transactions American Society of Agricultural Engineers*, 4: 62-64.

Dochinger, L.S. 1980. Interception of airborne particles by tree planting. *J. Environmental Quality*, 9: 265-268.

Dodd, C.K. Jr., Enge, K.M. and Stuart, J.N. 1989. Reptiles on highways in north-central Alabama, USA. *J. Herpetology*, 23: 197-200.

Dowler, R.C. and Swanson, G.A. 1982. High mortality of cedar Waxwings associated with highway plantings. *Wilson Bulletin*, 94: 602-604.

Drayton, B. and Primack, R.B. 1996. Plant species lost in an isolated conservation area in Metropolitan Boston from 1894 to 1993. *Conservation Biology*, 10: 30-39.

Dumont, A-G., Schneider, S. and Tille, M. 1999. *Conference Fauna and Traffic. Traffic Systems and Fauna Networks; Necessity for a New Approach. Proc. Lausanne Conference*. Ecole Polytechnic Federale de Lausanne.

Eaglin, G.S. and Hubert, W.A. 1993. The effects of logging and roads on substrate and trout in streams of the Medicine Bow national Forest, Wyoming. *North American J. Fisheries Management*, 13: 844-846.

ECMT. 1990. *Transport Policy and the Environment*. ECMT ministerial session. European Conference of Ministers in cooperation with OECD. OECD, Paris.

Edmunds, J. 1995. Head-on Collision 1995. *The Wildlife and Roads Report*. A report prepared for the Cheshire, Cumbria and Lancashire Wildlife Trusts, UK.

Ehmann, H. and Cogger, H. 1985. Australia's endangered herpetofauna: a review of criteria and policies pp. 435-447. In: G. Grigg, R. Shine and H. Ehmann (eds.). *Biology of Australasian Frogs and Reptiles*. Surrey Beatty & Sons and Royal Society of New South Wales, Sydney.

Environmental Resources Management. 1996. The significance of secondary effects from roads and road transport on nature conservation. *English Nature Research Reports* No. 178, Peterborough.

Erickson, P.A., Camagis, G. and Robbins, E.J. 1978. *Highways and Ecology: Impact Assessment and Mitigation*. National Technical Information Services, Virginia.

Evans, D.R. and Gates, J.E. 1997. Cowbird selection of breeding areas: the role of habitat and bird species abundance. *Wilson Bulletin*, 109: 470-480.

Eversham, B.C. and Telfer, M.G. 1994. Conservation value of roadside verges for stenotopic heathland Carabidae: corridors of refugia? *Biodiversity and Conservation*, 3: 538-545.

Evink, G.L. 1990. Wildlife crossings of Florida I-75. *Transportation Research Record*, 1279: 54-59.

Evink, G.L. 1998. Ecological highways pp. 253-257. In: Evink, G.L., Garrett, P., Zeigler, D. and Berry, J. (eds.). *Proc. International Conference on Wildlife Ecology and Transportation*, FL-ER-69-98, Florida Department of Transportation, Tallahassee, Florida.

Evink, G.L., Garrett, P., Zeigler, D. and Berry, J. 1996. Trends in addressing transportation related mortality. *Proc. Transportation Related Wildlife Mortality*

Seminar. State of Florida Department of Transportation Environmental Management Office. FL-ER-58-96.

Evink, G.L., Garrett, P., Zeigler, D. and Berry, J. 1998. *Proc. International Conference on Wildlife Ecology and Transportation*. FL-ER-69-98, Florida Department of Transportation, Tallahasse, Florida.

Extence, C.A. 1978. The effects of motorway construction on an urban stream. *Environmental Pollution*, 17: 245-252.

Fahrig, L., Pedlar, J.H., Pope, S.E., Taylor, P.D. and Wegner, J.F. 1995. Effect of road traffic on amphibian density. *Biological Conservation*, 73: 177-182.

FAO. 1989. Definition and scope of protective measures for roads pp. 1-12. In: *Watershed Management Field Manual—Road Design and Construction in Sensitive Watersheds*. FAO Conservation Guide 13/5. FAO, Rome.

Farmer, A.M. 1993. The effects of dust on vegetation—a review. *Environmental Pollution*, 79: 63-75.

Feitelson, E. and Papay, N. 1999. Sharing rights of way along inter-urban corridors: a spatial temporal and institutional analysis. *Transportation Research*, D, 4: 217-240.

Feldhamer, G.A., Gates, J.E., Harman, D.M., Loranger, A.J. and Dixon, K.R. 1986. Effects of instate fencing on white-tailed deer activity. *J. Wildlife Management*, 50: 497-503.

Feltwell, J. and Phillip, J. 1980. The natural history of the M40 motorway. *Transactions of the Kent Field Club*, 8, 101-115.

Ferris, C.R. 1979. Effects of Interstate 95 on breeding birds in northern Maine. *J. Wildlife Management*, 43: 421-427.

Finnis, R. 1960. Road casualties among birds. *Bird Study*, 7: 21-32.

Flanagan, J.T., Wade, K.J., Curie, A. and Curtis, D.J. 1980. The deposition of lead and zinc from traffic pollution on two roadside shrubs. *Environmental Pollution* (Series B), 1: 71-78.

Fletcher, T. and McMichael, A.J. (ed.). 1997. *Health at the crossroads: Transport Policy and Urban Health*. John Wiley and Sons, Chichester.

Fluckiger, W., Oertli, J.J. and Fluckiger-Keller, H. 1978. The effect of wind gusts on leaf growth and foliar water relations of Aspen. *Oecologia*, 34: 101-106.

Foner, H.A. 1987. Traffic lead pollution of some edible crops in Israel. *Science Total Environment*, 59: 309-315.

Fookes, P.G., Sweeney, M., Manby, C.N.D. and Martin, R.F. 1985. Geological and geotechnical engineering aspects of low-cost roads in mountainous terrain. *Engineering Geology*, 21: 1-152.

Forbes, B.C. 1995. Tundra disturbance studies, III: short term effects of aeolian sand and dust, Yamal region, northwest Siberia. *Environmental Conservation*, 22: 335-344.

Forman, R. and Alexander, L. 1998. Roads and their ecological effect. *Annual Review of Ecological Systematics*, 29: 207-231.

Forman, R. and Deblinger, R. 2000. The ecological road-effect zone of a Massachusetts (USA) suburban highway. *Conservation Biology*, 14: 36-46.

Forman, R.T.T. 1995. *Corridor Attributes and Roads. An Land Mosaics, the Ecology of Landscapes and Regions*. Cambridge University Press, Cambridge.

Forman, R.T.T. 2000. Estimate of the area affected ecologically by the road system in the United States. *Conservation Biology*, 14: 31-35.

Forman, R.T.T. and Hersperger, A.M. 1996. Road ecology and road density in different landscapes, with international planning and mitigation solutions pp. 1-22. In: Evink, G.L., Garrett, P., D. Zeigler, D. and Berry, J. (eds.). *Trends in Addressing Transportation Related Wildlife Mortality*. FL-ER-58-96. Florida Department of Transportation, Tallahassee.

Foster, M.L. and Humphrey, S.R. 1992. *Effectiveness of Wildlife Crossings in Reducing Animal/auto Collisions on Interstate 75, Big Cypress Swamp, Florida*. Florida Game and Freshwater Fish Commission, Report FL-ER-50-92.

Free, J.B., Gennard, D., Stevenson, J.H. and Williams, I.H. 1975. Beneficial insects present on a motorway verge. *Biological Conservation*, 8: 61-72.

Gates, J.E. and Gysel, L.W. 1978. Avian nest dispersion and fledging success in field-forest ecotones. *Ecology*, 59: 871-883.

Gerrard, S. 1995. Environmental Risk Management, pp. 296-316. In: T. O'Riordan, (ed.). *Environmental Science for Environmental Management*. Longman Scientific and Technical, Harlow.

Getz, L.L., Cole, F.R. and Gates, D.L. 1978. Interstate roadsides as dispersal routes for *Microtus pennsylvanicus*. *J. Ecology*, 59: 208-213.

Gilbert, O.L. 1989. *The Ecology of Urban Habitats*. Chapman and Hall, London.

Gilbert, M. 1998. The Australian partnership approach to protecting roadside habitats pp. 189-194. In: Evink, G.L., Garrett, P., Zeigler, D. and Berry, J. (eds.). *Proc. International Conference on Wildlife Ecology and Transportation*. FL-ER-69-98. Florida Department of Transportation, Tallahassee.

Gilpin, A. 1995. *Environmental impact assessment (EIA): Cutting Edge for the Twenty-first Century*. Cambridge University Press, Cambridge.

Gish, C.D. and Christensen, R.E. 1973. Cadmium, nickel, lead and zinc in earthworms from roadside soil. *Environmental Science and Technology*, 7: 1060-1062.

Gittins, P. 1983. Road casualties solve toad mysteries. *New Scientist* 97: 530-531.

Given, D.R. 1994. *Roadsides, Railway Margins and Waterways. Forgotten Natural Habitats*. North Canterbury Conservation Board. May 1994.

Gjessing, E., Lygren, E., Anderson, S., Berglind, L., Carlberg, G., Efraimsen, H., Kallqvist, T. and Martinsen, K. 1984. Acute toxicity and chemical characteristics of moderately polluted runoff from highways. *Science Total Environment*, 33: 225-232.

Glaister, S., Burnham, J., Stevens, H. and Travers, T. 1998. *Transport Policy in Britain*, MacMillan Press, Basingstoke.

Good, R. and Grenier, P. 1994. Some environmental impacts of recreation in the Australian Alps. *Australian Parks and Recreation*, 30: 20-26.

Gordon, D. 1991. *Steering a New Course: Transportation, Energy and the Environ-ment*. Island Press, Washington.

Greene, D.L., Jones, D.W. and Delucchi, M.A. (eds.). 1997. *The Full Costs and Benefits of Transportation: Contributions to Theory, Method and Measurement*. Springer, Berlin.

Greszta, J. 1982. Accumulation of heavy metals by certain tree species pp. 161-165. In: Bornkamm, R., Lee, J.A. and Seaward, M.R.D. (eds). *Urban Ecology—The Second European Ecological Symposium*, Berlin, 8-12 September, 1980, Blackwell Scientific Publications, Oxford.

Grue, C.E., O'Shea, T.J. and Hoffman, D.J. 1984. Concentrations and reproduction in highway-nesting barn swallows. *The Condor*, 86: 383-389.

Grue, C.E., Hoffman, D.J., Beyer, W.N., and Franson, L.P. 1986. Lead concentrations and reproductive success in European starlings *Sturnis vulgaris* nesting within highway roadside verges. *Environmental Pollution* (Series A), 42: 157-182.

Haeck, J., Hengeveld, R. and Turin, H. 1980. Colonization of road verges in three Dutch Polders by plants and ground beetles (Coleoptera: Carabidae). *Entomologia Generalis* 6 (2/4): 202-215.

Haigh, M.J. 1982. Road development and rural stress in India, Himalaya. *Nordia*, 16: 135-140.

Haigh, M.J. 1983. Is the Karakoram safe and stable? *Himalaya: Man and Nature*, 6: 5-6.

Haigh, M.J. 1984. Landslide prediction and highway maintenance in the Lesser Himalaya, India. *Z. Geomorphologie*. Suppl., 51: 17-38.

Haigh, M.J. 1994. Case 4: Deforestation in the Himalayas pp. 440-462. In N. Roberts Ed. *The Changing Global Environment*, Blackwell, Oxford.

Haigh, M.J., Rawat, J.S., Bartarya, S.K. and Rawat, M.S. 1993. Factors affecting landslide morphology along new highways in central Himalaya. *Trans., Japanese Geomorpological Union*, 14-2: 99-123.

Haigh, M.J., Rawat, J.S., Rawat, M.S., Bartarya, S.K. and Rai, S.P. 1995. Interactions between forest and landslide activity along new highways in the Kumaun Himalaya. *Forest Ecology and Management*, 78: 173-189.

Hamer, M. 1987. *Wheels Within Wheels: A Study of the Road Lobby*. Routledge and Kegan Paul, London.

Hamilton, R.S. and Harrison, R.M. (eds.) 1991. *Highway Pollution*. Studies in Environmental Science, 44. Elsevier, Amsterdam.

Hansen, E. 2000. *Orchid Fever: A Horticultural Tale of Love, Lust and Lunacy*. Methuen, London.

Harper-Lore, B. and Wilson, M. Eds. (2000). *Roadside use of native plants*. Washington, Island Press.

Harris, L.D. and Gallagher, P.B. 1989. New initiatives for wildlife conservation. The need for movement corridors pp. 11-34. In: G. Mackintosh (ed.). *Defense of Wildlife: Preserving Communities and Corridors*, Defenders of Wildlife, Washington.

Harris, L.D. and Silva-Lopez, G. 1992. Forest fragmentation and conservation of biological diversity pp. 197-237. In: Fiedler, P.L. and Jain, S.K. (eds.). *Conservation Biology*, Chapman & Hall, London.

Haskell, D.G. 2000. Effects of forest roads on Macroinvertebrate soil fauna of the southern Appalachian Mountains. *Conservation Biology*, 14: 57-63.

Hassel, J.H., van Ney, J.J. and Garling, Jr., D.L. 1980. Heavy metals in a stream ecosystem at sites near highways. *Trans. American Fisheries Society*, 109: 636-643.

Havlin, J. 1987. Motorways and birds. *Folia Zool.*, 36: 137-153.

Hellawell, J.M. 1988. Toxic substances in rivers and streams. *Environmental Pollution*, 50: 61-85.

Hill, D. and Hockin, D. 1992. Can roads be bird friendly? *Landscape Design*, February 1992, pp. 38-41.

Holzner, W., Kriechbaum, M., Kutzenberger, H. and Bohmer, K. 1989. Die Bedeutung der straaaenhegleiten – den flachen fur den naturschutz – naturnahe gestaltung und management. Bundesministerium fur Wirtschaftliche Angelegenheiten, Strassenforschung, 371, Vienna.

Hodgen, R. and Ford, C.D. 1985. The planning, design and construction of a bypass through an area of outstanding natural beauty. *Proc. Institute of Civil Engineers*, Part 1, 78: 1065-1083.

Hodkinson, D.J. and Thompson, K. 1997. Plant dispersal: the role of man. *J. Applied Ecology*, 34: 1484-1496.

Hodson, N. and Snow, D. 1965. The road deaths enquiry, 1960-61. *Bird Study*, 9: 90-99.

HMSO, 1994. Royal Commission on Environmental Pollution, 18th. Report, Transport and the Environment. London, HMSO.

Horner, R.R. and Mar, B.W. 1983. Guide for assessing water-quality impacts of highway options and maintenance. *Transportation Research Record* 948: 47-55.

Horner, R.R. and Mar, B.W. 1985. Assessing the impacts of operating highways on aquatic ecosystems. *Transportation Research Record*, 1017: 47-55.

Howell, J. 1999. *Roadside Bio-engineering*. HM Government, Nepal.

Howorth, L.S. 1991. Highway construction and wetland loss: mitigation banking programme in the southeastern United States. *The Environmental Professional*, 13: 139-144.

Hughey, L.M. 1941. Mammalian invasion via the highway. *J. Mammology*, 22: 383-385.

Hunt, A., Dickens, H.J. and Whelan, R.J. 1987. Movement of mammals through tunnels under railway lines. *Australian Zoologist*, 24: 89-93.

Hynson, J., Adamus, P., Tibbets, S. and Darnell, R. 1982. *Handbook for Protection of Fish and Wildlife from Construction of Farm and Forest Roads*. Report 82/18. U.S. Fish and Wildlife Service, Office of Biological Services.

IEA. 1995. *Guidelines for Baseline Ecological Assessment*. Institute of Environmental Assessment. E. & FN Spoon, London.

Illner, H. 1992. Effect of roads with heavy traffic on grey partridge (*Perdix perdix*) density. *Gibier Faune Sauvage*, 9: 467-480.

Isermann, K. 1977. A method to reduce contamination and uptake of lead by plants from car exhaust gases. *Environmental Pollution*, 12: 199-203.

IUCN 1996. *Tourism, Ecotourism and Protected Areas*. IUCN, The World Conserva-tion Union, Gland.

Jaarsma, C.F. and Langevelde, F. 1997. Right-of-way management and habitat fragmentation: an integral approach with the spatial concept of the traffic calmed rural area pp. 383-392. In: Williams, J.R., Goodrich-Mahoney, J.W., Wisniewski, J.R. and Wisniewski, J. (eds.). *Sixth International Symposium on Environmental Concerns in Rights-of-way Management*, Elsevier, Oxford.

Jackson, S. 1996. Underpass systems for amphibians pp. 224-227. In: Evink, G., Ziegler, D., Garrett, P., and Berry, J. (eds.). *Transportation and Wildlife: Reducing Wildlife Mortality and Improving Wildlife Passageways across Transportation Corridors*, Federal Highway Administration and Florida Depart-ment of Transportation. Conference Report No. FHWA-PD-96-041, Florida.

Jackson, S.D. and Griffin, C.R. 1998. Toward a practical strategy for mitigating highway impacts on wildlife, pp. 17-22. In: Evink, G.L., Zeigler, D., Garrett, P., and Berry, J. (eds.). *Proc. International Conference on Wildlife Ecology and Transportation*. FL-ER-69-98, Florida Department of Transportation, Tallahassee.

Janssen, A.A.A.W, Lenders, H.J.R. and Leuven, R.S.E.W. 1997. Technical state and maintenance of underpasses for badgers in the Netherlands pp. 362-366. In: Canters, K. (ed.). Habitat Fragmentation and Infrastructure. *Proc. International Conference 'Habitat Fragmentation, Infrastructure and the Role of Ecological Engineering'*, Maastricht, The Hague. Delft, Ministry of Transport, Public Works and Water Management.

Jefferies, D. 1975. Different activity patterns of male and female badgers (*Meles meles*) as shown by road mortality. *J. Zoology*, London, 177: 504-506.

Johnson, N.P. 1990. Nesting Bald Eagles (*Haliaeetus leucocephalus*) in urban areas of southeast Alaska: assessing highway construction and disturbance impacts. *Transportation Research Record*, 1279: 60-68.

Jones, J.A., Swanson, F.J., Wemple, B.C. and Snyder, K.U. 2000. Effects of roads on hydrology, geomorphology, and disturbance patches in stream networks. *Conservation Biology*, 14: 76-85.

Jonkers, D.A. and De Vries, G.W. 1977. *Verkeersslachtoffers onder de Fauna*. Nederlandse Vereniging tot Bescherming van Vogels, Zeist.

Kaczensky, P., Knauer, F., Huber, T., Jonozovic, M. and Adamic, M. 1996. The Ljubljana-Postojna highway—a deadly barrier for brown bears in Slovenia. *J. Wildlife Research*, 1: 263-267.

Kadlec, R.H. 1994. Wetlands for water polishing: free water surface wetlands pp. 335-508. In: W. Mitsch (ed.). Global Wetlands: *Old World and New.*, Elsevier Science, B.V.

Kaimowitz, D. 1996. Livestock and deforestation in Central America in the 1980s and 1990s: a policy perspective. *Ciencias-Veterinarias-Heredia*, 1-2: 113-161.

Kammerbauer, H., Selinger, H., Rommelt, R., Ziegler Jons A., Knoppik, D. and Hock, B. 1986. Toxic effects of exhaust emissions on Spruce *Picea abies* and their reduction by the catalytic converter. *Environmental Pollution* (Series A), 42: 133-142.

Kavtaradze, D.N., Nikolaeva, L.F., Porschneva, E.B. and Florova, N.B. 1999. *Automobile Roads in Ecological Systems*. Moscow State University, Moscow (In Russian).

Keals, N. and Majer, J.D. 1991. The conservation status of ant communities along the Wubin-Perenjori corridor pp. 387-393. In: Saunders, D.A. and Hobbs, R.J. (eds.). *Nature Conservation 2. The Role of Corridors*, Surrey Beatty & Sons, Chipping Norton.

Keller, J. and Lamprecht, R. 1995. Road dust as an indicator for air pollution transport and deposition: an application of SPOT imagery. *Remote Sensing of the Environment*, 54: 1-12.

Keller, V. and Pfister, H.P. 1997. Wildlife passages as a means of mitigating effects of habitat fragmentation by roads and railway lines pp. 70-80. In: Canters, K. (ed.). Habitat Fragmentation and Infrastructure. *Proc. International Conference 'Habitat Fragmentation, Infrastructure and the Role of Ecological Engineering'*, Maastricht, The Hague, Delft, Ministry of Transport, Public Works and Water Management.

Kerlinger, P. and Lein, M. 1988. Causes of mortality, fat condition and weights of wintering snowy owls. *J. Field Ornithology*, 59: 7-12.

Kerri, K.D., Racin, J.A. and Howell, R.B. 1985. Forecasting pollutant loads from highway runoff. *Transportation Research Record* 1017: 39-46.

King, C.M., Innes, J.G., Flux, M., Kimberley, M.O., Leathwick, J.R. and Williams, D.S. 1996. Distribution and abundance of small mammals in relation to habitat in Pureora Forest Park. New Zealand. *J. Ecology*, 20: 215-240.

King, J.G. and Tennyson, L.C. 1984. Alteration of stream flow characteristics following road construction in north central Idaho. *Water Resources Research*, 20: 1159-1163.

Knutson, R. 1987. Flattened Fauna: *A Field Guide to Common Animals of Roads, Streets and Highways*. Ten Speed Press, Berkely, CA.
Kober, W.W. and Kehler, S.E. 1987. An analysis of design features in mitigating highway construction impacts on streams. *Transportation Research Record*, 1127; 50-60.
Kraus, G.H.M. and Kaiser, H. 1977. Plant response to heavy metals and sulphur dioxide. *Environmental Pollution*, 12: 63-71.
Ksaibati, K., Cha, C.Y. and Plancher, H. 1995. A new bitumen modifier produced by co-retorting scrap tires and waste oils. *Road and Transport Research*, 4: 46-59.
Kupfer, J. 1996. Patterns and determinants of edge vegetation of a midwestern forest preserve. *Physical Geography*, 17: 62-76.
Kushlan, J.A. 1988. Conservation and management of the American crocodile. *Environmental Management*, 12: 777-790.
Laaksovirta, K., Olkkonen, H. and Alakuijala, P. 1976. Observations on the lead content of lichen and bark adjacent to a highway in southern Finland. *Environmental Pollution*, 11: 247-255.
Lagerwerff, J.V. and Specht, A.W. 1970. Contamination of roadside soil and vegetation with cadmium, nickel, lead and zinc. *Environmental Science and Technology*, 4: 583-586.
Lalo, J. 1987. The problem of road kill. *Amer. Forests* (September-October), pp. 50-52, 72.
Lamont, B.B., Rees, R.G., Witkowski, E.T.F. and Whitten, V.A. 1994. Comparative size, fecundity, and ecophysiology of roadside plants of *Banksia hookeriana*. J. *Applied Ecology*, 1: 137-144.
Langton, T.E.S. (ed.). 1989. *Amphibians and Roads. Proc. Road Tunnel Conference, Rendsburg*, January 1989. ACO Polymer Products Ltd., England.
Larwood, J.G. and Markham, D. 1995. *Roads and Geological Conservation. A Discussion Document*. English Nature, Peterborough.
Laursen, K. 1981. Birds on roadside verges and the effect of mowing on frequency and distribution. *Biological Conservation*, 20: 59-68.
Lausi, D. and Nimis, P.L. 1985. Roadside vegetation in boreal South Yukon and adjacent Alaska. *Phytocoenologia*, 13: 103-138.
Leedy, D.L. 1975. *Highway-wildlife relationships. Vol. 1: A State of the Art Report*. Ellicott City, MD, Urban Wildlife Research Centre Report FHWA-RD-76-4, Final Report.
Leedy, D.L. 1978. Highways and wildlife: implications for management pp. 364-383. In: *Classi-fication, Inventory, and Analysis of Fish and Wildlife Habitat, Proc. National Symposium. Biological Services Programme*. Fish and Wildlife Service, U.S. Dept. of the Interior.
Lelievre, M. and Serodes, J-B. 1995. A new approach for the identification of environmental issues at stake in a road project. *J. Environmental Management*, 44: 221-231.
Li, X., Wang, W., Li, F. and Deng, X. 1999. GIS based map overlay method for comparative assessment of road environmental impact. *Transportation Research*, D, 4: 147-158.
Lister, D.B. 1992. The argument for mitigation—case studies of impact mitigation concerning Anadromous Salmonid habitat. *American Fisheries Society Symposium*, 13: 115-124.

Lonsdale, W.M. and Lane, A.M. 1994. Tourist vehicles as vectors of weed seeds in Kakadu National Park, Northern Australia. *Biological Conservation*, 69: 277-283.

Lord, B.N. 1987. Nonpoint source pollution from highway storm-water runoff. *Science Total Environment*, 59: 437-446.

Lotschert, W. and Kohm, H-J. 1978. Characteristics of tree bark as an indicator in high-emission areas. II: Contents of heavy metals. *Oecologia*, 37: 121-132.

Lotz, M.A., Land, E.D. and Johnson, K.G. 1996. *Evaluation of S.R. 29 Wildlife Crossing*. State of Florida Department of Transportation, Final report study No. 7583, FL-ER-64-96.

Luker, M. and Montague, K. 1994. *Control of Pollution from Highway Drainage Discharges*, Construction Industry Research and Information Association, report 142 London.

MacDonald, I.A.W. and Frame, G.W. 1988. The invasion of introduced species in nature reserves in tropical savannas and dry woodlots. *Biological Conservation*, 44: 67-93.

MacDonald, I.A.W., Loope, L.L., Usher, M.B. and Hamann, O. 1989. Wildlife conservation and the invasion of nature reserves by introduced species: a global perspective pp. 215-255. In: Drake, J.A., Mooney, H.A., di Castri, F., Groves, R.H., Kruger, F.J., Rejmanek, M. and Williamson, M. (eds.). *Biological Invasions—A Global Perspective*, SCOPE & John Wiley & Sons, NY.

Maddison, D., Pearce, D., Johansson, Calthrop, E., Liman, T. and Verhoef, E. 1996. *The True Cost of Road Transport*. Earthscan Publ. Ltd., London.

Mader, H.J. 1984. Animal habitat isolation by roads and agricultural fields. *Biological Conservation*, 29: 81-96.

Mader, H.J., Schell, C. and Kornacker, P. 1990. Linear barriers to arthropod movements in the landscape. *Biological Conservation*, 54: 209-222.

Madsen, A.B. 1996. Otter (*Lutra lutra*) mortality in relation to traffic, and experience with newly established fauna passages at existing road bridges. *Lutra*, 39: 76-89.

Maestri, B. and Lord, B.L. 1987. Guide for mitigation of highway stormwater runoff pollution. *Science Total Environment*, 59: 467-476.

Malcolm, J.R. and Ray, J.C. 2000. Influence of timber extraction routes on central African small-mammal communities, forest structure, and tree diversity. *Conservation Biology*, 14: 1623-1638.

Mallick, S., Hocking, G. and Driessen, M. 1998. Road kills of the eastern barred bandicoot (*Perameles gunii*) in Tasmania: an index of abundance. *Wildlife Research*, 25: 139-145.

Maltby, L., Forrow, D.M., Boxall, A.B.A., Calow, P. and Betton, C.I. 1995. The effects of motorway runoff on freshwater ecosystems. I. Field Study. *Environmental Toxicology and Chemistry*, 14: 1079-1092.

Manderville, V. and King, D. 1995. *Vehicle/Wildlife Collisions*. New Brunswick Department of Transportation, Canada. 45 pp.

Manen, F.T., Coley, A.B. and Peltor, M.R. 1995. *Use of Interstate Passageways by Black Bears. Final Report*. Federal Highway Administration, USA, Report No. TN-RES1058.

Mansergh, I.M. and Scotts, D.J. 1989. Habitat continuity and social organisation of the mountain pygmy-possum restored by tunnel. *J. Wildlife Management*, 53: 701-707.

Marrs, R.H., Frost, A.J. and Lunnis, R.A. 1992. The effects of herbicide drift on semi-natural vegetation: the use of buffer zones to minimize risks. *Aspects of Applied Biology*, 29: 57-64.
Massemin, S., Maho, Y.L. and Handrich, Y. 1998. Seasonal pattern in age, sex and body condition of barn owls (*Tyto alba*) killed on motorways. *Ibis*, 140: 70-75.
May, S.A. and Norton, T.W. 1996. Influence of fragmentation and disturbance on the potential impact of feral predators on native fauna in Australian forest ecosystems. *C.S.I.R.O. Wildlife Research*, 23: 387-400.
McClure, H. 1951. An analysis of animal victims on Nebraska's highways. *J. Wildlife Management*, 15: 410-420.
McCreight, J.D. and Schroder, D.B. 1977. Cadmium, lead and nickel content of *Lycoperdon periatum* Pers. in a roadside environment. *Environmental Pollution*, 13: 265-268.
McLellan, B.N. and Shackleton, D.M. 1988. Grizzly Bears and resource-extraction industries: effects of roads on behaviour, habitat use and demography. *J. Applied Ecology*, 25: 451-460.
McNeill, A. 1996. Road construction and river pollution in south-west Scotland. *J. Institution of Water and Environmental Management*, 10: 175-182.
McRae, M. 1997. Road kill in Cameroon. *Natural History*, 106, 14 pp.
Michael, E.D. 1986. Use of roadside plantings by songbirds for nesting. *Transportation Research Record*, 1075: 19-20.
Morgan, E.L., Porak, W.E. and Arway, J.A. 1983. Controlling acidic-toxic metal leachates from Southern Appalachian construction slopes: mitigating stream damage. *Transportation Research Record*, 948: 10-16.
Morris, M.G., Thomas, J.A., Snazell, R.G., Pywell, R.F., Stevenson, M. and Webb, N.R. 1994. Recreation of early successional stages for threatened butterflies—an ecological engineering example. *J. Environmental Management*, 42: 119-135.
Morris, P. and Morris, M. 1988. Distribution and abundance of hedgehogs (*Erinaceus europaeus*) on New Zealand roads. *New Zealand J. Zoology*, 15: 491-498.
Motto, H.L. and Daines, R.H. 1979. Lead in soil and plants: its relationship to traffic volume and proximity to highways. *Environmental Science and Technology*, 4: 231-237.
Mowbray, A.Q. 1968. *Road to Ruin*. Lippincott, Philadelphia.
Mukherjee, A., Mukherjee, G., Banerji, U. and Mukhopadhyay, S.P. 1998. Occupational exposure of the traffic personnel of Calcutta to lead and carbon monoxide. *Pollution Research*, 17: 359-362.
Munguira, M.L. and Thomas, J.A. 1992. Use of road verges by butterflies and burnet moth populations, and the effect of roads on adult dispersal and mortality. *J. Applied Ecology*, 29, 316-329.
Murphy, E.C. and Dowding, J.E. 1994. Range and diet of stoats (*Mustela erminea*) in a New Zealand Beech Forest. *New Zealand J. Ecology*, 18: 1-18.
Murphy, S.M. and Curatolo, J.A. 1987. Activity budgets and movement rates of caribou encountering pipelines, roads, and traffic in northern Alaska. *Can. J. Zoology*, 65: 2483-2490.
Muskett, C.J. and Jones, M.P. 1980. The dispersal of lead, cadmium and nickel from motor vehicles and effects on roadside invertebrate macrofauna. *Environmental Pollution*, 23: 231-242.
Nankinov, D.N. and Todorov, N.M. 1983. Bird casualties on highways. *Sov. J. Ecol.*, 14: 288-293.

National Research Council, 1992. *Restoration of Aquatic Ecosystems: Science, Technology, and Public Policy*. National Academy Press, Washington DC, USA.

Nelson, R. and Holben, B. 1986. *Identifying deforestation in Brazil using Multiresolution Satellile data International J. Remote Sensing*, 7(3) pp. 429-448.

NEPA. 1970. The Environmental Policy Act of 1969. PL91-190, 91st Congress, S. 1075, Jan. 1. Washington, USA.

New Zealand House of Representatives, 1998. *Inquiry into the Environmental Effects of Road Transport. Interim report of the Transport and Environment Committee*. New Zealand House of Representatives, Wellington.

Newbey, B.J. and Newbey, K.N. 1987. Bird dynamics of Foster Road Reserve, near Ongerup Western Australia pp. 341-343. In: Saunders, D.A., Arnold, G.W., Burbridge, A.A. and Hopkins, A.J., (eds.). *Nature Conservation: the Role of Remnants of Native Vegetation*, Surrey Beatty and Sons Ltd., Chipping Norton, Australia.

Nieuwenhuizen, W. and van Apeldoorn, R.C. 1995. *Mammal Use of Faunal Passages on National Road A1 at Oldenzaal*. Ministry of Transport, Public works and Water Management, Report No. P-Dwww-95.737, Delft, The Netherlands.

Nilson, D. 1977. Roadside management and wetland development along North Dakota Highways. *North Dakota Outdoors*, 40: 23-25.

Norton, D.A. and Smith, M.S. 1999. Why might roadside mulgas be better mistle hosts? *Australian J. Ecology*, 24: 193-198.

Noss, R.F. 1987. Protecting natural areas in fragmented landscapes. Natural *Areas Journal*, 7: 2-13.

Noss, R.F., 1995. The ecological effects of roads or the road to destruction. (Unpublished report).

NRTEE 1997. *The Road to Sustainable Transportation in Canada*. National Round-table on Environment and Economy (NRTEE), Ottawa.

Oberts, G.L. 1986. Pollutants associated with sand and salt applied to roads in Minnesota. *Water Resources Bull.* 22: 479-483.

Odum, E.P. 1959. *Fundamentals of Ecology*, 2nd. ed. W.B. Saunders & Co., Philadelphia and London.

OECD. 1994. *Environmental Impact of Roads*. OECD Road Transport Research, Scientific Expert Group, OECD, Paris.

OECD 1997. *Road Transport Research—Outlook 2000—Thirteenth Anniversary*. Organisation for Economic Co-operation and Development, Paris.

O'Flaherty, C.A. 1997. *Transport Planning and traffic engineering*. London, Arnold.

Olthoff, T. 1986. Untersuchungen zur insektenfauna Hamburger strassenbaume. *Entomologische Mitteilungen aus dem Zoologischen Museum Hamburg*, 8: 213-229.

O'Neil, D.H., Robel, R.J. and Dayton, A.D. 1983. Lead contamination near Kansas highways: implications for wildlife enhancement programmes. *Wildlife Society Bull.*, 11: 152-160.

Owens, L.K. and James, R.W. 1991. Mitigation of traffic mortality of endangered brown pelicans on coastal bridges. *Transportation Research Record*, 1312, 3-14.

Oxley, D.J., Fenton, M.B. and Carmody, G.R. 1974. The effects of roads on populations of small mammals. *J. Applied Ecology*, 11: 51-59.

Parendes, L.A. and Jones, J.A. 2000. Role of light availability and dispersal in the exotic plant invasion along roads and streams in the H.J. Andrews experimental forest, Oregon. *Conservation Biology*, 14: 64-75.

Parizek, R.R. 1971. Impact of highways on the hydrogeologic environment pp. 151-199. In: Coates, D.R. (ed.). *Environmental Geomorphology and Landscape Conservation*, Hutchinson and Ross, Stroudsburg.
Perring, F.H. and Farrel, L. 1977. *British Red Data Books 1: Vascular Plants*. Royal Society for Promotion of Nature Conservation, Nettleham.
Port, G.R. and Thompson, J.R. 1980. Outbreaks of insect herbivores on plants along motorways in the U.K. *J. Applied Ecology*, 17: 649-656.
Port, G.R. and Spencer, H.J. 1987. Effects of roadside conditions on some Auchenorrhyncha pp. 7-11. *Proc. Sixth Auchenorrhyncha Meeting*, Torino.
Potteiger, M. and Purinton, J. 1998. *Landscape Narratives. Design Practices for Telling Stories*. John Wiley & Sons, Inc., NY.
Pratt, C.J. 1984. Design limits on pollution. *Science Total Environment*, 33: 161-170.
Pratt, J.M. and Coler, R.A. 1976. A procedure for the routine biological evaluation of urban runoff in small rivers. *Water Research*, 10: 1019-1025.
Przbylski, Z. 1979. The effects of automobile exhaust gases on the Arthropods of cultivated plants, meadows and orchards. *Environmental Pollution*, 19: 157-161.
Purseglove, J. 1989. *Taming the Flood. A History and Natural History of Rivers and Wetlands*. Oxford University Press, Oxford.
Quarles, H.D. III, Hanawalt, R.B. and Odum, W.E. 1974. Lead in small mammals, plants and soil at varying distances from a highway. *J. Applied Ecology*, 11: 937-949.
Queensland Department of Main Roads. 1997. *Roads in the Wet Tropics. Planning, Design, Construction, Maintenance and Operation Best-Practice Manual*. Document No. 66515. Gutteridge, Haskins & Davey Pty Ltd., Cairns.
Radford, T. 1916. *The Construction of Roads and Pavements*. McGraw-Hill Book Co., Inc., NY.
Ramsay, D. (ed.) 1994. *Roads and Nature Conservation. Guidance on Impacts, Mitigation and Enhancement*. English Nature, Peterborough.
Reed, D.F. 1981. Mule deer behaviour at a highway underpass exit. *J. Wildlife Management*, 45: 542-543.
Reed, D.F., Woodard, T.N. and Pojar, T.M. 1975. Behavioural responses of mule deer to a highway underpass. *J. Wildlife Management*, 39: 361-367.
Reed, R.A., Johnson-Barnard, J. and Baker, W.L. 1996. Contribution of roads to forest fragmentation in the Rocky Mountains. *Conservation Biology*, 10: 1098-1106.
Reeve, H.A. 1977. Evaluation of roadside verges. *Watsonia*, 11: 148-149.
Reid, J.W. and Bowles, I.A. 1997. Reducing the impacts of roads on tropical forests. *Environment*, 39: 10-13.
Reid, L.M. and Dunne, T. 1984. Sediment production from forest road surfaces. *Water Resources Research*, 20: 1753-1761.
Reid, R.A., Johnson-Barnard, J. and Baker, W.L. 1996. Contribution of roads to forest fragmentation in the Rocky Mountains. *Conservation Biology*, 10: 1098-1106.
Reijen, R. and Foppen, R. 1994. The effects of car traffic on breeding bird populations in woodland. I. Evidence of reduced habitat quality for willow warblers (*Phylloscopus trochilus*) breeding close to a highway. *J. Applied Ecology*, 31: 85-94.
Reijnen, M.J.S.M., Veenbass, G. and Foppen, R.P.B. 1995. *Predicting the effects of motorway traffic on breeding bird populations*. Road and Hydraulic Engineering Division & DLO-Institute for Forestry and Nature Research, Delft.

Reiter, P. and Sprenger, D. 1987. The used tire trade: a mechanism for the worldwide dispersal of container breeding mosquitoes. *J. American Mosquito Control Assoc.*, 3: 494-501.

Reuter, J.E., Djohan, T. and Goldman, C.R. 1992. The use of wetlands for nutrient removal from surface runoff in a cold climate region of California—results from a newly constructed wetland at Lake Tahoe. *J. Environmental Management*, 36: 35-53.

Rich, A.S., Dobkin, D.S. and Niles, L.J. 1994. Defining forest fragmentation by corridor width; the influence of narrow forest-dividing corridors on forest-nesting birds in southern New Jersey. *Conservation Biology*, 8: 1109-1121.

Richardson, J.R., Shore, R.F., Treweek, J.R. and Larkin, S.B.C. 1997. Are major roads a barrier to small animals? *J. Zoology*, London, 243: 840-846.

Riffell, S. 1999. Road mortality of dragonflies (Odonata) in a Great Lakes coastal wetland. *Great Lakes Entomologist*, 31: 63-73.

Roach, G.L. and Kirkpatrick, R.D. 1985. Wildlife use of roadside woody plantings in Indiana. *Transportation Research Record*, 1016: 11-15.

Rodriguez-Flores, M. and Rodriguez-Castellon, E. 1982. Lead and cadmium levels in soil and plants near highways and their correlation with traffic density. *Environmental Pollution* (Series B), 4: 281-290.

Rodwell, J.S. (ed.). 1991. *British Plant Communities, vol. 1. Woodlands and Scrub.* Cambridge University Press, Cambridge.

Romin, L.A. and Bissonette, J.A. 1996. Deer-vehicle collisions: status of state monitoring activities and mitigation efforts. *Wildlife Society Bulletin*, 24: 276-283.

Roof, J. and Wooding, J. 1996. *Evaluation of S.R. 46 Wildlife Crossing.* State of Florida Department of Transportation. Technical Report no. 54. FL-ER-61-96.

Rose, R.J. and Webb, N.R. 1994. The effects of temporary ballast roadways on heathland vegetation. *J. Applied Ecology*, 31: 642-650.

Royal Society for Natural Conservation. 1986 *Wild Flowers on the Verges: A Guide for Users and Managers.* RSNC, Lincoln, UK.

Ruediger, B. 1996. The relationship between rare carnivores and highways, pp. 1-12. In: Evink, G.L., Garrett, P., Zeigler, D. and Berry, J. (eds,). *Trends in Addressing Transportation Related Mortality, Proc. Transportation Related Wildlife Mortality Seminar.* State of Florida Department of Transportation, Environ-mental Management Office. FL-ER-58-96.

Salisbury, R.W. 1993. *The Fragmentation of Wildlife Conservation Areas by Linear Structures.* MSc. Thesis, Macquarie University, Sydney.

Salisbury, R.W. 1996. *A Design for Studying Edge Effects in Forests.* Working Paper 9606, Graduate School of Environment. Macquarie University, Sydney.

Salvig, J.C. 1991. Faunapassager i forbindelse med storre vejanlaeg. Enudredningsopgave udfort i samarbejde med Skov-og Naturstyrelsen. Denmark Miljoundersogelser.

Samways, M., Osborn, R. and Carliel, F. 1997. Effect of a highway on ant (Hymenoptera: Formicidae) species composition and abundance, with recommendation for roadside verge width. *Biodiversity and Conservation*, 6: 903-913.

Santelmann, M.V. and Gorham, E. 1988. The influence of airborne road dust on the chemistry of Sphagnum mosses. *J. Ecology*, 76: 1219-1231.

Sarkar, R.K., Banerjee, A. and Mukherji, S. 1986. Acceleration of peroxidase and catalase activities in leaves of wild dicotyledonous plants as an indication of automobile exhaust pollution. *Environmental Pollution* (Series A), 42: 289-295.

Sawyer, J. 2000. Off shore islands, mainland islands and traffic islands: New Zealand approaches to plant conservation. Unpubl. paper, Wellington Conservancy, Department of Conservation, Wellington, New Zealand.

Scanlon, P.F. 1987. Heavy metals in small mammals in roadside environments: implications for food chains. *Science Total Environment*, 59: 317-323.

Scanlon, P.F. 1991. Effects of highway pollutants upon terrestrial ecosystems pp. 281-291. In: Hamilton, R.S. and Harrison, R.M. (eds.). *Highway Pollution*, Elsevier, Amsterdam.

Schafer, J., Hannker, D., Eckhardt, J-D. and Stuben, D. 1998. Uptake of traffic-related heavy metals and platinum group elements (PGE) by plants. *Science Total Environment*, 215: 59-67.

Schaffers, A.P. 2000. *Ecology of Roadside Plant Communities*. Ph.D. thesis, Department Environmental Sciences, Wageningen University, The Netherlands.

Schmidly, D.J. and Wilkins, K.T. 1977. Composition of small mammal populations on a highway right-of-way in east Texas. *Texas State Department of Highways and Public Transportation*, Research Report 197-1F.

Schmidt, W. 1989. Plant dispersal by motor cars. *Vegetatio*, 80: 147-152.

Schmitt, R.J. and Osenberg, C.W. 1996. *Detecting Ecological Impacts: Concepts and Applications in Coastal Habitats*. Academic Press, NY.

Schonewald-Cox, C. and Buechner, M. 1992. Park protection and public roads pp. 373-395. In, Fiedler, P.L. and Jain, S.K. (eds.). *Conservation Biology*, Chapman & Hall, London.

Schullery, P. 1987. The problem of road kill. *American Forests*, September/October 93, 5 pp.

Scott, N.E. and Davison, A.W. 1985. The distribution and ecology of coastal species on roadsides. *Vegetatio*, 62: 433-440.

Seabrook, W.A. and Dettman, E.B. 1996. Roads as an activity corridor for cane toads in Australia. *J. Wildlife Management*, 60: 363-368.

Seargent, A.B. 1981. Road casualties of prairie ducks. *Wildlife Society Bull.*, 9: 65-69.

Servheen, C., Waller, J. and Kasworm, W. 1998. Fragmentation effects of high speed highways on Grizzly Bear populations shared between the United States and Canada pp. 97-103. *In Proc. International Conference on Wildlife Ecology and Transportation*, Florida, 1998, Florida Department of Transportation, Tallahassee.

Sheate, W.R. and Taylor, R.M. 1990. The effect of motorway development on adjacent woodland. *J. Environmental Management*, 31: 261-267.

Shelley, P.E., Driscoll, E.D., and Sartor, J.D. 1987. Probabilistic characterization of pollutant discharges from highway stormwater runoff. *Science Total Environment*, 59: 401-410.

Sherburne, J. 1985. Wildlife populations utilizing right-of-way habitat along interstate 95 in Northern Maine. *Transportation Research Record*, 1016: 16-20.

Simini, M. and Leone, I.A. 1986. Studies on the effects of de-icing salts on roadside trees. *Arboriculture J.*, 10: 221-231.

Simmons, J. 1938. *Feathers and Fur on the Turnpike*. Christopher Publ. House, Boston.

Singer, F.J., Langlitz, W.L. and Samuelson, E.C. 1985. Design and construction of highway underpasses used by Mountain Goats. *Transportation Research Record,* 1016: 6-10.

Singh, A.P. 2000. Designing railroads, highways and canals in protected areas to reduce men-elephant conflicts. *7th International Symposium, Environmental Concerns in Rights-of-way Management,* Calgary, Canada (in press).

Smith, D.J., 1995. *The Direct and Indirect Impacts of Highways on the Vertebrates of Payne's Prairie State Preserve.* University of Florida, Department of Wildlife Ecology and conservation. FL-ER-62-96.

Smith, D.L. and Lord, B.N. 1990. Highway water quality control—summary of 15 years of research. *Transportation Research Record.* 1279: 69-74.

Smith, D.S. 1993. Greenway case studies. In Smith, D.S. and Hellmund, P. (eds.). *Ecology of Greenways—Design and Function of Linear Conservation Areas.* Univ. Minnesota Press, Minneapolis.

Smith, D.S. and Hellmund, P. (eds.). 1993. *Ecology of Greenways—Design and Function of Linear Conservation Areas.* Univ. Minnesota Press, Minneapolis, pp. 161-208.

Smith, W.H. 1976. Lead contamination of the roadside ecosystem. *J. Air Pollution Control Association,* 26: 753-765.

Sorokovikova, N.V. 1990. The overall effect of automobile transportation on the natural environment. *Soviet Geography,* 31: 116-125.

Southerland, M.T. 1995. Conserving biological diversity in highway development projects. *Environmental Professional,* 17: 226-242.

Spellerberg, I.F. 1991. *Monitoring Ecological Change.* Cambridge University Press, Cambridge,

Spellerberg, I.F. 1992. *Evaluation and Assessment for Conservation. Ecological Guidelines for Determining Priorities for Nature Conservation.* Chapman and Hall, London.

Spellerberg, I.F. 1994. The biological content of environmental assessments. *Biologist,* 41: 126-128.

Spellerberg, I.F. 1998. Ecological effects of roads and traffic: a literature review. *Global Ecology and Biogeography Letters,* 7: 317-333.

Spellerberg, I.F. and Gaywood, M.J. 1993. Linear features: linear habitats and wildlife corridors. *English Nature Research Report No. 60.* English Nature, Peterborough.

Spellerberg, I.F. and Sawyer, J.W.D. 1996. Standards for biodiversity: a proposal based on biodiversity standards for forest plantations. *Biodiversity and Conservation,* 5: 447-459.

Spellerberg, I.F. and Sawyer, J.W.D. 1999. *An Introduction to Applied Biogeography.* Cambridge University Press, Cambridge.

Spencer, H.J. and Port, G.R. 1988. Effects of roadside conditions on plants and insects. II. Soil conditions. *J. Applied Ecology,* 25: 709-715.

Spencer, H.J., Scott, N.E., Port, G.R. and Davison, A.W. 1988. Effects of roadside conditions on plants and insects. I. Atmospheric conditions. *J. Applied Ecology,* 25: 699-707.

Stauch, C. 1998. Modelling the ecological effects of traffic on a regional scale using GIS pp. 533-537. In: Breuste, J., Feldmann, H. and Uhlmann, O. (eds.). *Urban Ecology,* Springer, Berlin.

Stauch, C. 2000. *GIS als entscheidungsunterstutzendes Werkzeug in der Verkehrsplanung am Beispiel von Flachenzerschneidung und Immissionsbelastung.* Diss. Institute for Landscape Planning and Ecology, Univ. Stuttgart. Published in digital format under http://elib.unistuttgart.de/opus/volltexte/2000/652.

Stotz, G. 1987. Investigations of the properties of the surface water run-off from Federal Highways in the FDR. *Science Total Environment,* 59: 329-337.

Stotz, G. 1990. Decontamination of highway surface runoff in the FRG. *Science Total Environment,* 93: 507-514.

Straker, A. 1998. Management of roads as biolinks and habitat zones in Australia pp. 181-188. In: Evink, G.L., Garrett, P., Zeigler D. and Berry J. (eds.). *Proc. International Conference on Wildlife Ecology and Transportation,* FL-ER-69-98. Florida Department of Transportation, Tallahassee, FL.

Susskind, L., McKearnan, S. and Thomas-Larmer, J. (eds.) 1999. *The Consensus Building Handbook: A Comprehensive Guide to Reaching Agreement.* Sage, California.

Swanson, C.S. and Loomis, J. 1998. Economic values associated with roaded and non-roaded recreation areas in the Pacific pp. 53-65. In: Evink, G.L., Garrett, P., Zeigler, D. and Berry, J. (eds.), *Proc. International Conference on Wildlife Ecology and Transportation,* FL-ER-69-98, Florida Department of Transporta-tion, Tallahassee, FL.

Swift, L.W. 1986. Filter strip widths for forest roads in the southern Appalachians. *Southern J. Applied Forestry,* 10: 27-34.

Swihart, R.K. and Slade, N.A. 1984. Road crossing in *Sigmodon bispidus* and *Microtus ochrogaster. J. Mammalogy,* 65: 357-360.

Tabor, R. 1974. Earthworms, crows, vibrations and motorways. *New Scientist,* 62: 482-483.

Tarrer, A.R., Whetstone, G.T. and Boylan, J.W. 1995. Impacts of environmental health and safety regulations on highway maintenance. In: *Maintenance Management. Proc. Seventh Maintenance Management Conference,* vol. 5, pp. 144-151. Transportation Research Board/National Research Council & National Academy Press, Washington, D.C.

Taylor, M.A.P., Woolley, J.E., Young, T.M. and Thompson-Clement, S.J. 1995. A modelling framework for including environmental impact analysis in transport planning studies: seeking prevention not cure. pp. 167–174. In: Sucharov, L.J. (ed.). *Urban Transport and the Environment.* Computational Mechanics Publ., Boston, MA..

Thaler, F., Bohmer, K., Kriechbaum, M. and Holzner, W. 1996. Vegetationsokologische forschungungen and strassenrandbiotopen. Bundesministerium fur Wirtschaftliche Angelegenheiten, Strassenforschung, 461, Vienna.

Theil, R.P. 1985. Relationship between road densities and wolf habitat sustainability in Wisconsin. *American Midland Naturalist,* 113: 404-407.

Thomas, J.A., Snazell, R.G. and Ward, L.K. In: Sherwood, B.R. (ed.). *Wildlife and Roads.* Linnean Society, London (in press).

Thompson, J.R., Rutter, A.J. and Ridout, P.S. 1986. The salinity of motorway soils. 1. Variation in time and between regions in the salinity of soils on central reserves. *J. Applied Ecology,* 23: 251-267.

Thorpe, H. 1978. The man-land relationship through time. In: Hawkes, J.J. (ed.). *Conservation and Agriculture,* pp. 17-44, Duckworth, London.

Thrasher, M.H. 1983. Highway impacts on wetlands: assessment, mitigation, and enhancement measures. *Transportation Research Record*, 948: 17-20.

Thull, J-P. 2000. *Management of Stock Truck Effluent Spillage in New Zealand*. Ph.D. Thesis, Lincoln University, NZ.

Tideman, J., Stacy, B. and Todd, P. 1996. The practical application of ecologically sustainable development concepts: analysis of two asphalt replacement processes. *Road and Transport Research*, 5: 14-26.

Timmins, S.M. and Williams, P.A. 1990. Reserve design and management for weed control pp. 133-138. In: Basset, C., Whitehouse, L.J. and Zabkiewicz, J.A. (eds.). *Alternatives to the Chemical Control of Weeds*. FRI Bulletin, 155. Ministry of Forestry, Forest Research Institute, Rotorua, New Zealand.

Timmins, S.M. and Williams, P.A. 1991. Weed numbers in New Zealand's forest and scrub reserves. New Zealand. *J. Ecology*, 15: 153-162.

Townsend, A.M. 1984. Effect of sodium chloride on tree seedlings in two potting media. *Environmental Pollution* (Series A), 34: 333-344.

Travis, M.D. and Tilsworth, T. 1986. Fish passage through Poplar Grove Creek Culvert, Alaska. *Transportation Research Record*, 1075: 21-26.

Treweek, J.R. 1999. *Ecological Impact Assessment*. Blackwell Science, Oxford.

Treweek, J.R., Thompson, S., Veitch, N. and Japp, C. 1993. Ecological assessment of proposed road developments: a review of environmental statements. *J. Environmental Planning and Management*, 36: 295-307.

Trimble, G.R. and Sartz, R.S. 1957. How far from a stream should a logging road be located? *J. Forestry*, May, 1957, pp. 339-341.

Trivedi, R.N., Kumar Binod, Mishra, S.K. and Singh, S.N. 1993. The studies of auto vehicular exhaust pollution in Patna (Bihar). *J. Mendel*, 10: 171-172.

Trombulak, S. and Frissell, C. 2000. Review of the ecological effects of roads on terrestrial and aquatic communities. *Conservation Biology*, 14: 18-30.

Tsunokawa, K. and Hoban, C. (eds.). 1997. *Roads and the Environment: A Handbook*. The World Bank, Herndon, VA

Thull, J-P. 2000. *Management of stock effluent spillage from trucks in New Zealand*. Ph.D. thesis, Lincoln University, NZ.

Tyser, R.W. and Worley, C.A. 1992. Alien flora in grasslands adjacent to road and trail corridors in Glacier National Park, Montana (USA). *Conservation Biology*, 6: 253-262.

Udevitz, M.S., Howard, C.A., Robel, R.J. and Carnutte, B. 1980. Lead contamination in insects and birds near an interstate highway, Kansas. *Environmental Entomology*, 9: 35-36.

UK: Dept. of Transport. 1989. *Roads for Prosperity*. HMSO, London.

UK: Dept. of Transport. 1992. *Assessing the Environmental Impact of Road Schemes*. HMSO, London.

UK, Dept. of Transport. 1994. *Royal Commission on Environmental Pollution, 18th Report*. HMSO, London.

UK: Parliamentary Commissioner for the Environment. 1990. *Audit of the future State Highway One. Route environmental impact report*, Vol. 1. Office of the Parliamentary Commissioner for the Environment, Wellington.

UK: Royal Commission on Environmental Pollution. 1994. *18th Report. Transport and the Environment*. HMSO. Cm 2674. London.

Ullmann, I. and Heindl, B. 1989. Geographical and ecological differentiation of roadside vegetation in temperate Europe. *Botanica Actu*, 102: 261-269.
Ullmann, I., Bannister, P. and Wilson, J.B. 1995. The vegetation of roadside verges with respect to environmental gradients in southern New Zealand. *J. Vegetation Science*, 6: 131-142.
Untermann, R.K. 1978. *Principles and Practices of Grading, Drainage and Road Alignment. An Ecologic Approach.* Reston Publ. Co., Reston, UK.
Usher, M.B. 1988. Biological invasions of nature reserves. A search for generalisations. *Biological Conservation*, 44: 119-135.
van Bohemen, H.D. 1995. Mitigation and compensation of habitat fragmentation caused by roads: strategy, objectives, and practical measures. *Transportation Research Record*, 1475, 133-137.
van Bohemen, H.D. 1998. Habitat fragmentation and roads: strategy, objectives and practical measures for mitigation and compensation pp. 574-578. In: Breuste, J., Feldmann, H. and Uhlmann, O. (eds.), *Urban ecology*, Springer, Berlin.
van Dyke, F.G., Brocke, R.H. and Shaw, H.G. 1986. Use of road track counts as indices of mountain lion presence. *J. Wildlife Management*, 50: 102-109.
van der Grift, E.A. and Kuijsters, R.M.J. 1998. Mitigation measures to reduce habitat fragmentation by railway lines in the Netherlands pp. 166-170. In: Evink, G.L., Garrett, P., Zeigler, D. and Berry, J. (eds.). *Proc. International Conference on Wildlife Ecology and Transportation*, FL-ER-69-98. Florida. Department of Transportation, Tallahassee, FL.
van der Zande, A.N. ter Keurs, W.J. and van der Weijden, W.J. 1980. The impact of roads on the densities of four bird species in an open field habitat—evidence of a long-distance effect. *Biological Conservation*, 18: 299-321.
Van Wee, B., Moll, H.C. and Dirks, J. 2000. Environmental impact of scrapping old cars. *Transportation Research*, D, 5: 137-143.
Veerabhadra Swamy, K.T. and Lokesh, K.S. 1993. Lead dispersion studies along highways. *Indian J. Environmental Health*, s35: 205-209.
Verboom, J. 1995. *Dispersal of Animals and Infrastructure. A Model Study: Summary.* Directorate-General for Public Works and Water Management, Road and Hydraulic Engineering Division. Delft, The Netherlands.
Voelk, F., Glitzer, I. and Woess, M. 2000. *Kostenreduktion bei Gruenbruecken durch rationellen Einsatz. 2. Zwischenbericht samt Checkliste Informationserfordernisse undverfuegbarkeit ueber jagdbare Wildtiere und habitate fuer Strassenplanung und Begutachtung in Oesterreich.* Auftrag des Oesterreichischen Bundesministeriums fuer Verkehr, Innovation und Technologie. Insitut fuer Wildbiologie und Jagdwirtschaft der Universitaet fuer Bodenkultur Wien.
Voelk, F., Glitzner, I., Woess, M., 2001: Kostenreduktion bei Gruenbruecken durch deren rationellen Einsatz. Kriterien - Indikatoren - Mindeststandards. *Bundesministerium fuer Verkehr, Innovation und Technologie.* Strassenforschung Helf 513. Wien. 211 Seiten.
Vos, C.C. 1997. Effects of road density: a case study of the moor frog pp. 93-97. In: Canters, K. (ed.). *Habitat Fragmentation and Infrastructure. Proc. International Conference 'Habitat Fragmentation, Infrastructure and the Role of Ecological Engineering'*, Maastricht, The Hague, Delft, Ministry of Transport, Public Works and Water Management.

Vos, C.C. 1999. A frog's eye view of the landscape: quantifying connectivity for fragmented amphibian populations. *Scientific Contributions* 18. Institute for Forestry and Nature Research, Wageningen.

Vos, C.C. and Chardon, J.P. 1998. Effects of habitat fragmentation and road density on the distribution pattern of the Moor Frog *Rana arvalis*. *J. Applied Ecology*, 35: 44-56.

Wace, N. 1977. Assessment of dispersal of plant species—the car borne flora in Canberra pp. 167-186. In: Anderson, D. (ed.). *Exotic Species in Australia—their establishment and success*, vol. 10, Proc. Ecological Society of Australia, Adelaide.

Wade, K.J., Flanagan, J.T., Currie, A. and Curtis, D.J. 1980. Roadside gradients of lead and zinc concentrations in surface dwelling invertebrates. *Environmental Pollution* (Series B), 1: 87-93.

Walder, B. 1998. About Wildlands CPR, pp. 234-235. In: Evink, G.L., Garrett, P., Zeigler, D. and Berry, J. (eds.). *Proc. International Conference on Wildlife Ecology and Transportation*, FL-ER-69-98. Florida Department of Transporta-tion, Tallahassee, FL.

Walker, D.A. and Everett, K.R. 1987. Road dust and its environmental impact on Alaskan Taiga and tundra. *Arctic and Alpine Research*, 19: 479-489.

Ward, A.L. 1982. Mule deer behaviour in relation to fencing and underpasses on Interstate 80 in Wyoming. *Transportation Research Record*, 859: 8-13.

Ward, N.I. 1990. Multielement contamination of British motorway environments. *Science Total Environment*, 93: 393-401.

Ward, N.I., Brooks, R.R. and Reeves, R.D. 1974. Effect of lead from motorway vehicle exhausts on trees along a major thoroughfare in Palmerston North, New Zealand. *Environmental Pollution*, 6: 149-158.

Ward, N.I., Reeves, R.D. and Brookes, R.R. 1975. Lead in soil and vegetation along a New Zealand State Highway with low traffic volume. *Environmental Pollution*, 9: 243-251.

Ward, N.I., Roberts, E. and Brooks, R.R. 1979. Seasonal variation in the lead content of soils and pasture species adjacent to a New Zealand highway carrying medium density traffic. New Zealand. *J. Experimental Agriculture*, 7: 347-351.

Warner, R.E. 1985. Demography and movements of free-ranging domestic cats in rural Illinois. *J. Wildlife Management*, 49: 340-346.

Warner, R.E. 1992. Nest ecology of grassland passerines on road rights-of-way in central Illinois. *Biological Conservation*, 59: 1-7.

Warren, R.S. and Birch, P. 1987. Heavy metal levels in atmospheric particulates, roadside dust and soil along a major urban highway. *Science Total Environment*, 59: 253-256.

Washington, C. 2000. *Pukeko road death at Otukaikino Reserve*. MSc. thesis, Lincoln University, NZ.

Watkins, L.H. 1981. *Environmental Impact of Roads and Traffic*. Applied Science Publ., London.

Way, J.M. 1977. Roadside verges and conservation in Britain: a review. *Biological Conservation*, 12: 65-74.

Webb, R.H. and Wiltshire, H.G. (eds.). 1983. *Environmental Effects of Off-road Vehicles: Impacts and Management in Arid Regions*. Springer-Verlag, NY.

Wells, M., Langton, T., Garland, L. and Wilson, D. 1996. The value of motorway verges for reptiles—a case study pp. 174-181. In: Foster J. and Gent, T. (eds.).

Reptile Survey Methods, Proc. Seminar, November 1995 at the Zoological Society London. English Nature Science No. 27. English Nature, Peterborough.
Wells, T.C.E. and Bayfield, N.G. 1990. *Wildflower Swards for Trunk Roads and Motorway Landscaping*. Institute of Terrestrial Ecology, Abbotts Ripton, Huntingdon, England.
Weste, G. 1977. Future forests—to be or not to be. *Victoria Resources*, 19: 26-27.
Wester, L. and Juvik, J.O. 1983. Roadside plant communities on Mauna Loa, Hawaii. *J. Biogeography*, 10: 307-316.
Westing, A.H. 1969. Plants and salt in the roadside environment. *Phytopathology*, 59: 1174-1181.
Wheeler, G.L. and Rolfe, G.L. 1979. The relationship between daily traffic volume and the distribution of lead in roadside soil and vegetation. *Environmental Pollution*, 18: 265-274.
Wilcox, B.A. and Murphy, D.D. 1985. Conservation strategy: the effects of fragmentation on extinction. *American Naturalist*, 125: 879-887.
Wilkie, D., Shaw, E., Rotberg, F., Morelli, G. and Auzel, P. 2000. Roads, development, and conservation in the Congo Basin. *Conservation Biology*, 14: 1614-1622.
Williams, P.A. and Buxton, R.P. 1995. Aspects of the ecology of two species of Passiflora (*P. mollissima* (Kunth) L. Bailey and *P. pinnatistipula* Cav.) as weeds in South Island, New Zealand. *New Zealand J. Botany*, 33: 315-323.
Williams-Linera, G. 1990. Vegetation structure and environmental conditions of forest edges in Panama. *J. Ecology*, 78: 356-373.
Wiliamson, L. 1980. Reflectors reduce deer-auto collisions. *Outdoor News Bull.*, 34: 2.
Wilson, D. 1977. Roadside management and wetland development along North Dakota highways. *North Dakota Outdoors*, 40: 23-25.
Wilson, J.B., Rapson, G.L., Sykes, M.T., Watkins, A.I. and Williams, P.A. 1992. Distributions and climatic correlations of some exotic species along roadsides in South Island, New Zealand. *J. Biogeography*, 19: 183-194.
Witmar, G.W. and de Calseta, D.S. 1985. Effects of forest roads on habitat use by Roosevelt elk. *Northwest Science*, 59: 122-125.
Wong, M.H., Cheung, Y.H. and Wong, W.C. 1984. Effects of roadside germination and root growth of *Brassica chinensis* and *B. parachinensis*. *Science Total Environment*, 33: 87-102.
Yahner, R.H. and Mahan, C.G. 1997. Effects of logging roads on depredation of artificial ground nests in forested landscape. *Wildlife Society Bull.*, 25: 158-162.
Yanes, M., Velasco, J.M. and Suarez, F. 1995. Permeability of roads and railways to vertebrates: the importance of culverts. *Biological Conservation*, 71: 217-222.
Yapp, W.B. 1973. Ecological evaluation of a linear landscape. *Biological Conservation*, 5: 45-47.
Young, A. and Mitchell, N. 1994. Microclimate and vegetation edge effects in a fragmented podocarp-broadleaf forest in New Zealand. *Biological Conservation*, 67: 63-72.
Young, A.G. and Clarke, G.M. 2000. *Genetics, Demography and Viability of Fragmented Populations*. Cambridge University Press, Cambridge.
Younkin, L.M. and Connelly, G.B. 1981. Prediction of storm-induced sediment yield from highway construction. *Transportation Research Record*, 832: 1-6.
Yousef, Y.A., Baker, D.M. and Hvitved-Jacobsen, T. 1996. Modelling and impact of metal accumulation in bottom sediments of wet ponds. *Science Total Environment*, 189/190: 349-354.

Zuckermann, W. 1991. *End of the Road. From World Car Crises to Sustainable Transportation.* Chelsea Green Publ. Co., VT, USA.

Zwaenepoel, A. 1997. Floristic impoverishment by changing unimproved roads into metalled roads pp. 127-137. In: Canters K. (ed.). *Habitat Fragmentation and Infrastructure, Proc. International Conference 'Habitat Fragmentation, Infrastructure and the role of Ecological Engineering'. Maastricht,* The Hague, Delft, Ministry of Transport, Public Works and Water Management.

Index

A

Animal warning devices 133
Archaeological 178
Archaeological 15
Artificial lighting 104
Asean Highway Project 4

B

Beneficial insects 53
Best practical environmental option 145
Biogeographical 90
Biogeographical regions 162
Biogeography 173
Biogeography 46, 78
Biological community 172
Biological surveys and inventories 47
Bitumen fumes 28
Bridges 131, 138
Buffer zones 61, 137

C

Campaigns 15
Cartoons 9
Cheshire Roadside Verge Survey 47, 48, 49, 50, 51, 65
Conflicts 143
Connectivity zones 90
Conservation banking 159
Consultation 182
Corridors 81, 140
Cost-benefit analysis 11, 137
Costs and benefits 165
Costs and benefits 87
Costs of mitigation of adverse effects 135
Culvert design 154

D

De-icer agents 110
Decommissioned roads 28
'Diversive' fragmentation 75
Drains 103

E

EC Directive 85/337 148
Ecoducts 42, 88
Ecological 45
Ecological effects 36, 44
Ecological evaluation 65
Ecological Impact Assessment (EcIA) 41
Ecological monitoring 182
Ecological restoration 156
Ecological restoration 45, 87
Ecological succession 76
Ecological surveys 185
Economic Co-operation and Development (OECD) 148
Ecosystem 38
Ecotone 172
Ecotone 85
Effects during construction 36
Effects from upgrading 36
Effects when roads are operational 36
Endangered 46
Environmental costs 13
Environmental Impact Assessment (EIA) 174
Environmental Impact Assessments 166
Environmental Impact Assessments (EIAs) 41
Environmental Policy Act of 1969 145
Erosion 103
Erosion 98

Erosion and sediment deposition 25
European Commission Directorate for General Transp 92

F

Fences 160
Films 9
Fires 29
Forest removal 102
Forest roads 25
Forest Service in the USA 152

G

Green bridges 42
Greenways 85

H

Habitat 172
Habitat 48, 52, 54, 57, 63, 64, 67
Habitat fragmentation 40
Habitat restoration 64
Habitats 39, 46, 187
Heavy metals 105
Herbicides and pesticides 112

I

Impact on tourism 33
Impervious surfaces 174
Incidental effects of roads 34
Incremental and cumulative 36
Incremental and ramifying effects 33
Infra Eco Network Europe (IENE) 92
Interested parties 143, 183
International Conference on Wildlife Ecology and T 22
Invasive plant species 26

K

Kerbs 103

L

LANDSCAPING 26
Landscaping 34
Legislation 144
Linear 39

Linear habitat 169
Linear habitats 84, 161
Linear nature reserves 165
Litter 112
Locomotive Act of 1865 7
Logging 103
Logging roads 102
Long-distance disturbance effects 104
Long-distance effects 98

M

MANAGING 57, 63
Microclimate 76, 78
Migration 79
Monitoring 189

N

National Ecological Network 92
National Environmental Policy Act (NEPA) 150
Nature conservation 34
Network of protected areas 73
Nitrogen deposition 108
Noise 177
Noise 102, 104, 123

O

OECD 11
Off-road vehicles 18
Off-road vehicles 99
Organisation for Economic Co-operation and Develop 9

P

Pathogenic fungi 83
Planning and assessments 155
Policies 154
Pollutants 40, 98, 165
Pollution 29, 40, 103, 109, 138, 155
Protected species 186

R

Rare species 65
Rarest plants 50
Reducing pollution 162
Resource Management Act (RMA) 150

Restoration 46, 87
Road accidents 118
Road construction 28
Road cuttings 101
Road design 101
Road dust 105
Road networks 33
Road policy 9
Road Tunnel Conference 129
Roadside nature reserves 161
Roadside vegetation 48
Roadside verge mowing 61
Roadside verges 27, 45, 46, 47, 48, 49, 50, 56, 57, 59, 64, 65
Roadside verges serve as refuges 53
Royal Society for Natural Conservation (RSNC) 60
Runoff 109

S

Sediment 103
Sediment load 109
Simmons Society 129
Snowmobiles 220
Social costs 13
Social issues 29
Soil compaction 102, 156
Spatial decision support systems 86
Stock truck effluents 112
Structures and utilities associated with roads and 32
Structures associated with roads 56
Sustainable transport policy 11

T

Tarmac Warriors 8
The Blues Highway 8

Toad tunnel 131
Traffic islands 87
Traffic noise 104
Traffic regulations 160
Traffic wind gusts 105
Tree theft 102
Tunnels 138
Tyre dumps 18

U

Underpasses 87, 131, 160
Urban environment 43
US Forest Service 150

V

VERGES 57, 60
Verges 168
Vibration 123

W

Water run 174
Water runoff 98
Weed 78, 83, 102
Weeds 51, 52
Weeds 78
Wetlands 109, 138, 140
Wilderness areas 136
Wildland CPR 150
Wildlands Centre for Preventing Roads (CPR) 19
Wildlife bridges 87, 160
Wildlife corridors 85, 169
Wildlife corridors 64
World Bank 4, 29, 142, 177
Wrecker yards 33